DISTANT
WANDERERS

DISTANT
WANDERERS

The Search for Planets Beyond the Solar System

BRUCE DORMINEY

COPERNICUS BOOKS
an imprint of springer-verlag

© 2010 Springer-Verlag New York, Inc.

Published in the United States by Copernicus Books,
an imprint of Springer-Verlag New York, Inc.
A member of BertelsmannSpringer Science+Business Media GmbH

Copernicus Books
37 East 7th Street
New York, NY 10003
www.copernicusbooks.com

Library of Congress Cataloging-in-Publication Data
Dorminey, Bruce
    Distant wanderers : the search for planets beyond the solar system / Bruce Dorminey.
        p. cm.
    Includes bibliographical references and index.

    1. Extrasolar planets. I. Title.
    QB820.D67 2001
    523—dc21   2001047132
ISBN 978-1-4419-2872-6
Manufactured in the United States of America.
Printed on acid-free paper.

9 8 7 6 5 4 3 2 1

To my parents, for their never-ending love and encouragement

For a thousand years in your sight are like a day that has just gone by, or like a watch in the night.

Psalm 90:4

Preface

Due to the efforts of a new global breed of planet hunters who move across time zones sharing ideas and technology with unprecedented ease, the number of planets known to circle other stars is now well over 60, and counting. We live in an era in which the detection of extrasolar planets has ceased to be a novelty and, instead, has become a challenge to the planetary theorists and theoretical astrophysicists attempting to explain them. Even during this book's two-and-a-half-year gestation period, I've been astounded not just by the frenetic pace of discovery, but by the mind-boggling variety of what astronomers have found. The discoveries range from protoplanetary disks to full-fledged planetary systems, but few, if any, appear to resemble our own Solar System.

The planets that have been discovered thus far seem to be gaseous giants that for the most part are in very close orbits around their host (or parent) stars. Some have orbital periods of only a few days. It's as if Jupiter, the most massive planet in our Solar System, were suddenly displaced from its nearly 12-year orbit around our Sun into an orbit that allowed it to circle our star in less than a week. The discovery of such planets presents theorists with many challenges, not the least of which is determining how the planets and planetary systems might actually have formed. In fact, these startling findings have required astronomers to rethink the very definition of the word "planet." Heretofore a planet (not including smaller objects such as asteroids, comets, and meteoroids) was loosely defined as any body that orbits a star but does not generate its own light, and generally has a maximum mass only a few times that of Jupiter. Given the latest data, this definition will have to remain very much in flux.

Detecting extra-solar planets is first and foremost a testament to the skill, hard work, and technological prowess of the world's astronomical community. Over the span of the twentieth century, technological developments enabled astronomers to expand their observations to cover almost the entire electromagnetic spectrum. Astronomers greatly benefited from the development of both larger-aperture optical telescopes and the power of computer processing to crunch their newly acquired data. The development of radio astronomy after World War II allowed for the combination of signals from more than one radio telescope. This technique of signal combination was later incorporated into optical telescopes and is a key part of present and future planet-detection technologies. Within the next 40 years, astronomers hope to be able to image and resolve surface details of Earth-like planets from both the ground and space. Progress has been and will continue to be staggering. Within the last hundred years, we will have gone from the 100-inch (2.5-meter) Hooker telescope atop Mount Wilson in California, to the Overwhelmingly Large Telescope (OWL), a 100-meter ground-based optical telescope now being planned by the European Southern Observatory that may see completion in 2016.

As the technology advances and the number of observed extra-solar planets increases, so will the number of questions. Already, there has been much debate over whether the "planets" that have been indirectly detected are actually planets. Besides forcing us to reconsider our definitions, this new information will likely have us asking, time and again, some of the fundamental questions of astronomy: How do planets form? How did our own Earth and Solar System form? When and why did Earth become habitable? Obviously, these are questions for which we still have—and may forever have—only incomplete answers. But the march of new technologies cannot be underestimated, for they will likely revolutionize our understanding of the Milky Way galaxy and our role in it.

Toward the end of the book, I speculate on some of the tantalizing questions that may be at least partially answered in the next few decades. Given our new tools, we may soon be able to directly image and remotely characterize the nature of many extra-solar planets. From distances light years away, will we detect vegetation? Will we find a pattern to what types of stars develop habitable planetary systems? What are the prospects of finding intelligent life around other Sun-like stars? From there, the questions only get more complex and involve the nature of our very existence: Why do we live in a Universe clumped into galaxies, with stars, circled by bodies we term planets, where the cycles of life and death as we understand them seem to play such a significant role? Does life on Earth, in some small way, mirror the cycles of birth and death in our own Sun and other stars like it? And, finally, are we, as a species, instinctively driven to leave Earth and our Solar System to avoid the inevitable destruction brought on by our own dying star?

The sheer number of stars that may produce planetary systems and the growing number of planets that we now know circle other stars are daunting, to say the least. Yet astronomy is singular in the fact that almost like no other science or subject, the more one knows, the more one appreciates not only the long hours of research being done nightly at observatories around the world, but also our Solar System's own history from protostellar nebula to protoplanetary disk to the formation of our own habitable planet. A night at an observatory is a humbling experience, not only because of the technologies and calculations involved in navigating the skies overhead, but because of the size and scope of what *is* overhead.

ACKNOWLEDGMENTS

One's raw material is only as good as one's access to the principal parties and relevant subject matter. Thus, I am thankful to everyone who played a role in providing information related to this subject matter, whether in the form of a formal interview, an E-mail, or a phone call. I am grateful to conference organizers for allowing me to sit in on their proceedings, and to the observatories for allowing me to get a closer look at their astronomers at work. Because at least half of my time was spent sifting through archives and periodicals trying to synthesize this moving target of a subject into an accessible and comprehensive form, *Distant Wanderers* would not have been possible without the use of the extensive archives of the Institut d'Astrophysique de Paris. Thanks to the institute's Roger Ferlet and its former librarian, Annick Benoist, and colleagues, for their assistance during the many hours I spent there. Thanks to Gregory Shelton and Brenda Corbin at the U.S. Naval Observatory library in Washington, D.C., for fielding my phone calls and E-mails and providing other needed reference materials. I also wish to thank the Naval Observatory's Steven Dick, Brian Mason, and Ralph Gaume for answering additional queries. Thanks to Maria López at the European Southern Observatory's Santiago de Chile Library for her help during my visit there, Marie-Angèle Lemoine at the ESA Headquarters' Paris library, the library at the Observatoire de Paris, and the ESA/ESTEC documentation center. Thanks to NASA, the European Space Agency, the European Southern Observatory, Sandra Legout at the Pasteur Institute, the Cerro Tololo Inter-American Observatory, Guy Webster and Jane Platt at the Jet Propulsion Laboratory, the Observatoire de Haute-Provence, Geneva Observatory, Colleen Gino at the VLA, Gemini North Observatory, W. M. Keck Observatory, Steward Observatory at the University of Arizona, and Phil Schewe at the American Institute of Physics.

I thank Franco Bonacina, Barbara Kennedy, Martin Kessler, Elaine MacAuliffe, Hernán Julio, Peter Michaud, Jorge Ianiszweski, Catherine Broomfield, Felipe Mac-Auliffe, Fabio Favata, Michael Mumma, James Kaler, Gordon Walker, Greg Henry, Ragbir Bhathal, Hal

McAlister, George Gatewood, Geoffrey Marcy, Didier Queloz, Antoine Labeyrie, Stephen Maran, Robert Sanders, Seth Shostak, Hans-Hermann Heyer, Valérie Muller, Edo Berger, David Buckley, David Cohn, Eric Feigelson, Sabine Frink, Tristan Guillot, Sandy Leggett, Jay Melosh, Claire Moutou, Richard Muller, Joan Najita, Rendong Nan, Maria Rosa Zapatero Osorio, Alan Penny, Deane Peterson, Jeff Pier, Andreas Quirrenbach, John Richer, Ted von Hippel, Barbara Welther, Lee Anne Willson, and everyone I interviewed.

I would especially like to thank Michel Mayor for initially inspiring my interest in this sub-ject, my agent, Robert Kirby, for his part in shepherding this book since its inception, and to his assistant, Catherine Cameron. Thanks to Rüdiger Gebauer at Springer-Verlag New York, John Watson at Springer-Verlag London, and former Copernicus editors Jonathan Cobb and Jerry Lyons for their part in giving this book a home.

Thanks to my editor, Janice Borzendowski, for all her diligence and many suggestions, Paul Farrell, Editor-in-Chief at Copernicus Books, for all his suggestions and perseverance in seeing this book into print, Assistant Editor, Anna Painter, for her suggestions during the final days of editing, and Production Editor, Mareike Paessler, for her keen sense of organization, suggestions, and much hard work. I also thank Copernicus Art Director Jordan Rosenblum and Designer Evan Schoninger for their role in the book's production.

Special thanks to everyone who offered solace during this process: my wonderful parents, Tina and Leroy, my brother Blair, my friends Eric Leibovitz, Lise Belanger, Avery Gliz-Kane, Caroline Nahoum, Marjorie Borradaile Bouillot, Halim Berretima, Louise Dowling, Brian Spence, and, finally, Martine Darroux for her enduring love and patience.

Bruce Dorminey
Paris, July 2001

brucedorminey@hotmail.com

CHAPTER 1
Cauldrons of Creation

On a clear night in the Northern Hemisphere, cruising the interstate highways across the Great Plains of the U.S., or ambling the foredecks of the old car ferries that still run the channel between the U.K. and France, the naked eye can discern some 5,000 stars. For many, stars are vague, never-shifting points of light, merely part of a handsomely crafted celestial backdrop for terrestrial tourist destinations. Or, at most, in this age of global satellite navigational systems, stars remain a crude and outmoded direction-finding reference. Yet we are inextricably linked to the stars. The very ground on which we stand is a by-product of their formation. And so are we. If not for stars, there would be no planets, no plants, no animals, no glass and steel, no Internet, and no iron-enriched vitamins to help us make it through the day.

In the beginning, there was hydrogen, helium, small amounts of a lithium isotope, and very little else. Just how this cosmic genesis cooled and expanded after the Big Bang in a way that allowed for the formation of molecular clouds of dust and gas remains a mystery. But we do know that carbon, oxygen, nitrogen, calcium, sulfur, silicon, iron, and gold all originate in stars. It took several generations of stars—massive gas bodies in which, through the process of thermonuclear fusion, hydrogen is converted to helium—to create enough metals to enable the formation of terrestrial-type planets. (Astronomers refer to any element that has an atomic weight heavier than helium as a metal.) And any hope we have of finding intelligent life elsewhere in the Universe, if it does exist, is also tied to these burning balls of hydrogen.

Consider how far we've come. It was less than 80 years ago that Edwin Hubble first proved that our own galaxy was only one of many. Until then, the consensus had been that the Milky Way galaxy constituted the whole observable Universe. We now know that is hardly the case. At last count, the Hubble Space Telescope has revealed there are closer to 50 billion galaxies in the Universe—and there could be double or even triple that number. Our own Milky Way galaxy alone contains an estimated 200 billion stars, and some astronomers estimate that the true number may be as many as 400 billion.

Planets are something else altogether. Six of the nine circling our own star (the Sun) are visible without binoculars or a telescope. But aside from the planet we call home, Mercury, Venus, Mars, Jupiter, and Saturn are as often as not confused with bright background stars. And for most of us, the speculation ends there, for very little seems capable of prodding us out of our Earth-centric complacency. Few people jump out of bed every morning in a dither over the fact that our Milky Way galaxy is 130,000 light years across.

But that will change. Before the end of this new century, every schoolchild will know for certain how many planets circle nearby stars; whether Earth-like planets are a rare anomaly, or, as one extra-solar planet hunter puts it, "extraterrestrial civilizations are buzzing from planet to planet like bumblebees hopping from flower to flower." We will have sent probes to the nearest stars, and we will have a definitive understanding of the size and scope of the observable Universe. There will be large permanent observatories on the far side of the Moon, and space-based observatories at the edge of our Solar System. Yet for now, time remains our nemesis, an inherently frustrating and all too relative construct. Time frames for planet formation, like other astrophysical processes, far outstrip our capacity for comprehension. For an astronomer waiting for a star to move across the arc of the sky, the passage of time requires an interminable amount of patience.

OUR LOCAL GALACTIC NEIGHBORHOOD

Some 400 years before Hubble, our two nearest dwarf galaxies had been spotted and mistakenly deemed "clouds." Portuguese explorer Ferdinand Magellan gets credit for circumnavigating the globe and for discovering the Large and Small Magellanic Clouds. In fact, he did neither. After taking a westerly route around the tip of South America on an ill-fated quest to reach Asia's Spice Islands, he was killed in what we now call the Philippines.

Only one of his two ships limped on to complete the circumnavigation, returning to Spain in 1522. But even before Magellan, the "Cape Clouds," as previous navigators had dubbed them, had been used as guideposts in the approach to southern Africa's Cape of Good Hope. Set off in a relatively dark part of the sky, the two dwarf galaxies are clearly visible without binoculars. They lie a little more than 150,000 light years away in the constellations Dorado and Tucana and are thought to orbit our own galaxy perpendicular to the plane of the Milky Way. They are important because they are the first extragalactic objects to be spotted and noted, even if Magellan and his predecessors had no concept of what constituted a galaxy. Their early discovery, and Hubble's later explanation of galaxies as "islands in the Universe," marks as much of a turning point in our own view of the cosmos as when we finally realized that Earth circles the Sun, not the other way around.[1]

Tracking Galactic Rotation

But no matter the epoch, no matter the observer, one fact remains: it's still a wild and wooly Universe out there. We lie in what can be thought of as the outer suburbs of one of billions of spiral galaxies, tucked away between two of an estimated four of the Milky Way galaxy's spiral arms. As the late Carl Sagan frequently noted, at first glance it seems that we circle an ordinary star in an ordinary galaxy. Our galaxy is not even one of the biggest. In a recent survey of 17 spiral galaxies, the Milky Way was found to be, at most, only average in size.

While most of the youngest stars are confined to the spiral arms, at 5 billion years old, our Sun is approximately halfway between the Carina Sagittarius and Perseus arms, slightly closer to the Sagittarius arm, at a distance of roughly 1,000 parsecs, or 3,260 light years.

NGC 1288, a spiral galaxy not unlike our own, lies some 300 million light years away in the southern constellation of Fornax. Its diameter is almost twice the size of our own Milky Way galaxy. (European Southern Observatory.)

PARSEC
A parsec is the astronomical unit for a distance of 3.26 light years; one light year, the distance
traveled by a beam of light (or other electromagnetic radiation) in a year, is roughly 9.46 trillion
kilometers. Our own Solar System is less than one light day across. By comparison, the nearest star in
our "neighborhood" is Proxima Centauri, which is part of the Alpha Centauri triple star system, 1.3
parsecs away.

In late 1999, astronomers at Leiden University in the Netherlands identified and tracked a dozen ancient stars that invaded the Milky Way some 10 billion years ago—leftover remnants from a dwarf galaxy of some 30 million stars. Their discovery gives credence to the idea that our own galaxy formed from a conglomeration of many dwarf galaxies, and that it is probably still forming. Clouds of high-velocity gas, mostly hydrogen, continue to rain down from the halo of the Milky Way toward its center, nurturing the enrichment and recycling process that is key to star formation. With four spiral arms extending from a long central flat disk, what we view as the Milky Way (referred to as the celestial river by the ancient Chinese) is best seen from May to September at 30 degrees south latitude (which bisects Argentina and Chile, cuts across most of Polynesia, all of southern Australia, South Africa, and a large portion of the southern Indian Ocean).

We know that our Solar System spun down from the gravitational collapse of a cold cloud of gas (mainly hydrogen) and dust into a protostellar disk. It is just as likely that our own galaxy collapsed into a flat disk from a spherical collection of hydrogen clouds. The same gravitational processes that caused the collapse of the Milky Way also created angular momentum. To imagine this event, visualize a figure skater going into a final spin at the end of a performance: she pulls in her arms to speed up. At the same time, her center of gravity spins down.

Our Solar System rotates due to angular momentum; moreover, our whole galaxy has momentum that is left over from its formation, which causes the spiral arms to continue to rotate around the center of the galaxy, some 26,000 light years, or 8,000 parsecs, away. And the Sun, like everything else in the Milky Way, is circling what is almost certainly a black hole.

The black hole at the hub of our galaxy is Sagittarius A* (pronounced "A star"), and we, here on Earth, circle it at a rate of once every 225 million years. Mark Reid, a radio astronomer at the Harvard Smithsonian Center for Astrophysics in Cambridge, Massachusetts, led a team in 1999 that conducted the first direct measurement of the angular rotation of the Milky Way. "Assuming that Sagittarius A* is at the center," says Reid, "we measure where it lies on the sky relative to some very distant quasars, which are at an infinite distance in the background. The black hole appears to move across the sky along the galactic plane, or Milky Way, which is perfectly explainable as the galaxy rotating. You can simulate this by looking at a wall in front of you: put your finger out at arm's length, and as you move your head from side to side, you will see your fingertip move with respect to the wall. The wall represents the distant quasars, and your fingertip is the black hole, the center of the Milky Way. Your eye is the Solar System, and as the whole Solar System spins around the Milky Way, you see this thing move. That's the entire measurement. You can sort of see the effect in 10 days, at least in the east-west direction. But in a month you could detect this with the VLBA [Very Long Baseline Array]."[2]

BLACK HOLE
Frequently mythologized in popular science fiction as wormhole entrances to alternative universes or distant parts of our Universe, black holes are, in fact, simply the endpoint of any star, roughly five to eight solar masses or larger, that has run of out fuel and collapsed onto itself. All nearby matter (including photons of visible light) will ultimately succumb to the gravitational field surrounding the hyperdense mass of a black hole. (Theoretical physicist Stephen Hawking has proposed the exception that black holes may emit "virtual" radiation—radiation that takes its energy from a black hole for an instant before it self-annihilates in a quantum state of particle/antiparticle production.) Most galaxies, including our own Milky Way, are now thought to have supermassive black holes at their centers, many of which could be made up of thousands of collapsed stars.

In order to "look" into space, Reid and his colleagues rely on the VLBA, a network of 10 radio telescopes spread over 5,000 miles, from Mauna Kea on the big island of Hawaii to St. Croix in the U.S. Virgin Islands. Operated by the National Science Foundation from the VLBA's operations center in Socorro, New Mexico, it was some 20 years ago that the VLBA first spotted Sagittarius A* as a black hole.

Yet the work done by Reid and his team would not have been possible if not for an astute observation made in 1932 by Karl Jansky, a young radio engineer who was studying static on a new transatlantic telephone link. Jansky noticed that much of the interference he was experiencing came not from thunderstorms or the Sun, but from the center of our galaxy. This realization marked the beginning of radio astronomy.[3]

BEGINNING OF THE END
Stars, then, are the seed and the cradle of all life, and galaxies are their incubators. But what triggers the formation of stars? Theorists believe supernovae, in part. Supernovae mark the beginning of the very end for a massive star. Stars normally maintain their equilibrium by balancing between the gravitational inward "pull" caused by the density of the star's own mass, and the outward "push" caused by the thermonuclear fusion of hydrogen atoms into helium. In massive stars that are on the verge of "going supernova," the source of hydrogen has been depleted to such an extent that the star's inner core has nothing left to burn; therefore, it literally collapses onto itself. What remains of the star then convulses in a violent explosion. During and immediately following this violent process, the supernova creates many of the heavy metals necessary to build terrestrial-type planets, like those in our Solar System. In a supernova explosion, metal-rich knots of ejected matter are hurled hundreds of thousands of kilometers outward from the star. The shock wave from the explosion heats the matter to such high temperatures that remnants of the supernova form rings of glowing gas. All that's left of the star's former self—aside from supernova detritus—is a supercompact neutron star core.

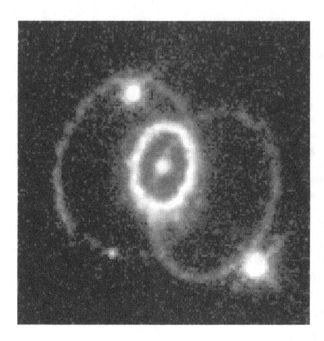

Rings from Supernova 1987A as seen in visible light by the Hubble Space Telescope in early 1994. (Dr. Christopher Burrows, ESA/STScI, and NASA.)

We still cannot fully comprehend the brute force and destructiveness of supernovae. We do know that without them, we certainly wouldn't be here, because only supernovae get hot enough to create iron. And even more fundamentally, supernovae may play a direct role in star formation. Although the conventional star-formation paradigm consists of collapsing clouds of nebular hydrogen, it is now believed that supernovae continually facilitate the process with shock waves that create superbubbles of hot gas that roll across the galaxy. Such superbubbles cool and fragment as they go, creating dark molecular clouds, which in turn become dense star-forming regions called clusters. Observations by the European Space Agency's (ESA's) Infrared Space Observatory (ISO) found one such young cluster in the constellation of Serpens. ESA calculated that the cluster has a local star density of over 400 stars per square parsec.

Supernovae occur on average only once every 50 years in our own galaxy. The first recorded observation of a supernova was made by the Chinese on July 4, 1054, who simply identified it as a "guest" star. The explosion of the star, about ten times the mass of the Sun, was so bright that it could be seen in broad daylight for three weeks. But it wasn't until 1942 that Dutch astronomer Jan Oort and his colleague Nicholas Mayall figured out that the visual remnants left over from the explosion matched the position of the Crab Nebula, some 6,500 light years away in the Taurus constellation. (Due to its unusual shape, the Crab Nebula had been spotted and named a hundred years earlier.)[4]

Perhaps the best observed instance of a supernova in the last century was that of Supernova 1987A, first seen on February 23, 1987, after Sanduleak −69 degrees 202 (known as

The Crab Nebula, a supernova remnant, and its rapidly spinning pulsar at the center, as observed in the X-ray spectrum by NASA's Chandra X-ray Observatory. Surrounded by rings of high-energy particles, which have been flung outward from the pulsar, its inner ring is estimated to be more than 1,000 times the diameter of our Solar System. (NASA/CXC/SAO.)

a "blue supergiant"—among the hottest, most massive and short-lived types of stars known) exploded in the Large Magellanic Cloud. 1987A became the first "naked-eye" supernova since 1604 and offered astronomers the first direct confirmation that heavy elements are produced in supernovae. Although carbon, nitrogen, and oxygen can all be created in midsized stars, iron is made only in the cores of stars that are soon to go supernova, and are at least eight times more massive than the Sun. "Iron is actually formed during the last stages of stellar evolution before the star blows up," explains John Hughes, an astrophysicist at Rutgers University in New Jersey.[5] Likewise, elements even heavier than iron, such as uranium and plutonium, can be formed only in the expanding waves of a supernova's blast.

During its August 1999 check-out and activation phase, NASA's $1.5 billion Earth-orbiting Chandra X-ray Observatory made observations of Cassiopeia A (Cas A), a supernova remnant 3,300 parsecs away. After studying the data from Cas A, Hughes concluded that gaseous clumps of silicon, sulfur, and iron had been expelled from the deepest part of the star, where temperatures reach up to 30 million degrees Kelvin.

"Our Sun is burning hydrogen to helium in its core in its stable phase, which lasts billions of years," says Hughes. "But when a very massive star gets to the last stages of stellar evolution, basically it's burning silicon to iron in its core. Iron has a peculiar nuclear feature, which is that when you try to fuse two iron nuclei together, you don't gain energy, you take energy away. When you fuse two hydrogen atoms you gain a little bit of energy. Energy gets released. But when two iron atoms combine, it's an endothermic reaction: it absorbs energy. So when iron dominates the core, there is no pressure left to stop the star's collapse, and the star becomes unstable. The iron in our Sun would have come from many previous generations of supernova remnants. Supernova remnants live for only a short period of time as distinctive objects. A hundred thousand years is a very old supernova remnant; beyond that, the motions in the galaxy destroy them."

Lying between the northern constellations of Cepheus and Cassiopeia, Cas A blew up over 9,000 years ago, and ever since, its outer envelope of superheated gas has been expanding at a rate of 800 kilometers per second. Its shell is now wider than the distance from here to the nearest star. Nevertheless, the iron formed in Cas A could very well end up thousands of parsecs away, in another one of our galaxy's spiral arms, forming new stars, whose circumstellar formation disks will eventually produce iron-rich planets like Earth.

PLANETS AROUND DEATH STARS

The first planets detected outside our Solar System orbit a pulsar, the radio-emitting neutron star that is left over after a supernova explodes. Neutron stars are so dense that if someone were having tea on the surface, a single lump of sugar would weigh a hundred million tons. Magnetically channeled radiation emitted from their poles makes pulsars seem to pulse as they spin. For example, the pulsar that formed from the core of the supernova at the center of the Crab Nebula is spinning at a rate of 30 times a second. Like that of other pulsars, its timing is so regular that it can rival the best atomic clocks. Futurists have even suggested using pulsars

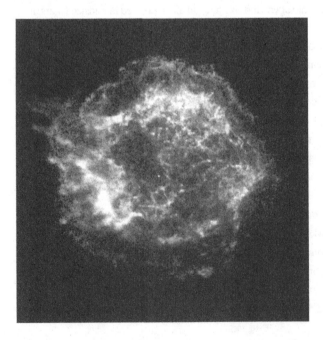

Cassiopeia A, as observed here in the X-ray spectrum, is a supernova remnant surrounded by iron-forming superheated gas rings. It is believed that in time this iron will eventually make up new terrestrial-type planets. (NASA/CXC/SAO.)

as aids in interstellar navigation, since each would have a distinctive pulse, or radio beacon signature, making it a reliable radio "lighthouse."

But planets around pulsars brought me to Bonn, Germany, and the Pulsar 2000 conference, an international meeting of some of the best radio astronomers in the world. I had scheduled an interview with Alexander Wolszczan, the man who had detected the first planetary system outside our Solar System, a notable achievement even though the planets he had discovered remain lifeless, pulsar-irradiated rocks. And I would also have a chance to meet and interview Jocelyn Bell Burnell, who, with guidance from her thesis adviser, discovered the first pulsar when she was a third-year doctoral candidate at Cambridge—proving that often science makes leaps when least expected. Who could have imagined that little more than 30 years ago, a student-built radio telescope made up of 2,000 dipole antennas, a thousand posts, and some 190 kilometers of wire and cable would stumble upon a whole new class of astronomical object?

Late in the spring of 1967, Bell Burnell was simply looking to spot quasars, which at the time were considered very exotic objects. (The name "quasar" is short for "quasi-stellar object," revealing just how tentative an understanding astronomers initially had of these objects.) Central to her search was a relatively new technique that employed radio telescopes to search for interplanetary scintillation, or diffracted radio waves. This technique had been recommended to Bell Burnell by Antony Hewish, her thesis adviser, as a good way to look for quasars. Bell Burnell's observations on the new radio telescope began in July of the same year. Her job was to analyze 30-meter paper rolls of the telescope's observations, which on average were made once every four days on four parts of the sky. The objective was to try to distinguish between what looked like ordinary interference, be it terrestrial or otherwise, and what looked like interplanetary scintillation from quasars.

Quasars, we now know, can lie at distances approaching the edge of the observable Universe. By August 6, Bell Burnell had spotted some peculiar interference that she termed "scruff." The emissions did not fit the typical profile she had previously observed from quasars, nor did they look like "normal" terrestrial interference. So she decided to continue to monitor this unusual interference to see what happened. To spread out the signal wavelengths in order to read their amplitude fluctuations more clearly, she sped up the chart paper. She soon realized that the scruff had appeared on the data charts before, and seemed to have come from the same part of the sky. It pulsed once every one-and-a-third seconds, like clockwork. By December, Bell Burnell had found another such source. As she told me at the conference in Bonn, by Christmas 1967, she and Hewish were jokingly referring to the signals as LGM (for little green men) 1–2, because they could find nothing natural about them. And they certainly weren't man-made.[6]

"I set up accurate timing of the pulses, and did it for about a month from Christmas of 1967 to January 1968," says Hewish, who was also at the conference. "I seriously didn't know how to handle the pulse if it turned out to be an intelligent signal. We decided that whatever the hell we did, [we would] not let the press know because they would overrun the place, and we wouldn't get any science done. Beyond that, we decided between ourselves that we would have to contact the Royal Society, and first of all convince them that the signals we had were most likely intelligence. We treated this as a bit of a joke, but we thought it could have been an intelligent beacon or something."[7] Hewish began making observations to check for a Doppler shift, in case it was an intelligent signal. He believed that if it were, it would most likely have come from a technology-bearing civilization on a planet orbiting a normal star not unlike our own. And if that were the case, Hewish would also be able to detect this Doppler wavelength-shifting of the signal as the pulse moved toward and away from Earth, along our line of sight, in essence mimicking the effect of a signal emanating from a planet in rotation around its star. This effect is one anyone can hear simply by listening to tonal changes from a passing car.

DOPPLER EFFECT

The Doppler effect is an observed change in the frequency of a wave due to the relative motion between a wave's source and its observer. To a pedestrian (the observer) standing still alongside an otherwise empty highway, the tonal pitch of a passing car (the source) traveling at a steady rate changes from higher to lower as the car passes. As the car approaches, the pedestrian hears the sound waves from the car as compressed, which causes a higher pitch. Then, once the car passes and recedes, its sound waves, from the observer's standpoint, are elongated, causing a lower pitch. In fact, there has been little or no change in the frequency of the sound waves produced by the car. What has changed, however, is what the observer perceives. The Doppler effect is based solely on our perceptions, and it operates with many kinds of wave phenomena, including all manner of electromagnetic radiation. Both radio and optical astronomers use the Doppler effect every day to track an object's relative velocity toward or away from them along their own line of sight, making it an essential tool in planet hunting.

Hewish did not detect any Doppler-shifting in the pulses, so he concluded that these various pulsing sources he and Bell Burnell continued to find over different parts of the sky, though indeed bizarre, were all within our galaxy and all natural, whatever they were. Pulsars had been theorized to exist since the 1930s, but it took a methodical year-long process of elimination until the astronomical community as a whole was totally comfortable with the present conclusion that indeed these were pulsating radio stars.

Bell Burnell went on to complete her thesis on quasars, but the pulsar revelation was given mention only in the appendix. Hewish, however, went on to win a 1974 Nobel Prize for his role in the discovery. Because Bell Burnell was only a doctoral candidate at the time of the initial discovery, convention dictated that she not share in the Nobel. But, through their collaboration, they made an amazing discovery using what is today regarded as crude 1960s technology. In a public lecture she delivered at the Bonn conference, Bell Burnell passed out envelopes inside each of which was a single sheet of paper with the note: "In picking up this piece of paper you have used a million times more energy than a radio telescope receives from all the known pulsars in a year." Even though these pulsating radio sources are faint and difficult to detect, today, more than a 1,000 pulsars have been cataloged. The fact that Alexander

Wolszczan was able to delineate a whole system of planets around one of them is even more remarkable.

Planets around Pulsars

Like Bell Burnell, who didn't set out to find pulsars, Wolszczan didn't set out to find planets around a pulsar. In 1990, he was simply searching for rare millisecond pulsars, pulsars that are spinning at literally hundreds of revolutions per second, in observations using the world's largest radio/radar telescope near Arecibo, Puerto Rico. At the time, Arecibo's steerable antenna was mostly out of commission while it underwent repairs, but Wolszczan, then of Cornell University, had applied for, and been granted, telescope observing time to chase millisecond pulsars in new regions of the sky.

How this telescope came to be built in a natural basin near Arecibo is a story that's equal parts politics and science. In 1958, William Gordon, then a professor of electrical engineering at Cornell University, initiated the idea for such a telescope, to be used in large part for studies of the Earth's atmosphere. In 1960, he raised $9 million to fund his project from the U.S. Department of Defense, which at the time (the height of the Cold War) was also very interested in conducting radar studies of Earth's ionosphere in order to find ways to improve strategic communications.

The natural mountaintop basin near Arecibo was chosen for the site because its curvature was perfectly suited to the telescope's 305-meter-diameter parabolic shape. The bowl of the telescope remains fixed, while a steerable antenna, attached to three pylons overhead, moves across the telescope. This gives it the means to fix and focus on its targets. Since its opening, the Arecibo Observatory has studied all manner of radio sources, including PSR 1257 +12, the millisecond pulsar around which Wolszczan would discover three terrestrial-type planets.

The Arecibo radio/radar telescope in Puerto Rico remains the world's largest single-dish radio telescope and most powerful radar. Its observations provided data that allowed Alexander Wolszczan to detect the first known planets to circle another star, even if that star is a dying pulsar. (Photo by David Parker, 1997/Science Photo Library.)

In most cases, the fast rate of rotation of these millisecond pulsars is caused by the spin-up from matter being drawn from a stellar "companion." Most millisecond pulsars are the product of two stars in orbit around each other; typically, one is massive and the other much smaller. The more massive star burns all its fuel much faster and evolves very quickly. Eventually, its core collapses into that of a neutron star, and most of its matter is blasted away as it "goes supernova." If the two stars survive as a double star, or binary, system, the neutron star usually evolves into a radio pulsar, which will emit pulses for millions of years. But it too will lose its energy and stop working like a pulsar, to eventually become just another dead neutron star orbiting a main-sequence star. Within several million, even billions, of years, its less massive normal companion star will gradually run out of hydrogen, lose hold of its matter, and enter what is known as a "red giant" phase. When that happens, the dead but gravitationally potent neutron star begins cannibalizing what's left of its more ordinary dying partner. That causes the neutron star to spin up into a "feeding frenzy," all the while rejuvenating itself as a pulsar as it whirls around its red giant companion. The highly magnetic, extremely dense, and therefore gravitationally potent pulsar literally begins drawing matter off its dying companion, which in turn manifests in an energetic spin-up of the radio pulsar—which in some cases can also emit X-rays. The more material that is pulled from its companion, the faster the pulsar's rate of rotation, sometimes reaching a rate of several hundred times per second. In this stable phase, the rotation rates of these radio and X-ray pulsars are again as precise as the best clocks on Earth.

Timing Pulsars

Wolszczan's strategy was simply to allow Earth's rotation to "point" Arecibo toward his target areas and see what would come up. He found PSR 1257 +12 in June of 1990.

> ### CELESTIAL LATITUDE AND LONGITUDE
> *PSR is an abbreviation for pulsar, and 1257 marks the hours and minutes of Right Ascension, the celestial equivalent to longitude, in this case, 12 hours and 57 minutes. The +12 refers to celestial latitude (or declination), with positive or negative notations marking the number of degrees above or below the celestial equator. PSR 1257 +12 lies about 600 parsecs away in the Virgo Constellation; it is heading down toward the plane of the galaxy at a rate of 300 kilometers per second. Unlike most millisecond pulsars, whose pulses rotate at rates of hundreds of times per second, PSR 1257 +12 is believed to be a solitary pulsar without any stellar companion. Wolszczan started timing it in January 1991.*

"I realized that this pulsar didn't time right," says Wolszczan, recounting the tale of PSR 1257 +12 during a break at the Bonn conference. "There was this extra variability that was hard to explain. At that time, we only knew of four millisecond pulsars. The common thinking was that millisecond pulsars were rock-solid clocks. The pulses go in a straight line with spikes, and we measure each pulse when it arrives. Then we integrate many such pulses and sum up all the changes in the pulsar period. If our prediction of how a pulsar period changes over months or years is correct, then we can predict the arrival time of pulses in the future. But it requires that we measure the pulse arrival time with an accuracy that is at least comparable to or better than atomic clocks."[8]

A pulsar's timing accuracy, or period, can change because it is losing energy and slowing; or, if it has planets, it will show variations in its wavelength amplitudes that are more oscillatory, indicating that the planet is moving toward or away from us along our line of sight. If the pulsar is simply slowing its rotation rate due to energy loss, then variations in its pulse timing will be what are known as "secular," meaning that the pulsar's signal amplitudes will move gradually and steadily in the same direction over time. Andrew Lyne, a radio astronomer at Jodrell Bank Observatory at the University of Manchester, U.K., notes that observing such microsecond variations in pulsar timing is equivalent to someone moving a clock 300 meters toward or away from you along your line of sight. As it moves closer, the clock will appear to be running early. As it moves away, the clock will appear to be running late. The millisecond pulsars in question are 600 to 700 parsecs away. And if they are orbiting each other in what is known as a "double neutron star binary system" (essentially two pulsars in rapid orbit around each other), their orbits are roughly a million kilometers across. "One microsecond," says Lyne, "is how long it takes radio waves to travel 300 meters. So we can measure their position accurately to about 300 meters." That's less than a lap around an Olympic-sized running track, meaning measurements at an astonishing degree of accuracy from distances of 600 to 700 parsecs.

Based on the size of the perturbations manifested in the radio timing of the pulsar itself, pulsar astronomers can extrapolate the mass of the object causing the perturbation to within a very small margin of error. However, a normal star in orbit around a pulsar would cause a perturbation a thousand times greater than what the astronomers would expect from an orbiting terrestrial-sized planet. So they can easily spot the difference between that of an ordinary binary pulsar system and that of a pulsar being circled by planets. Even so, to predict when a signal will arrive from a pulsar that is harboring planets requires an enormous amount of computer modeling.

Astronomers must measure the pulse arrival times and then try to make the models fit. Wolszczan began looking for a periodicity (that is, a repetitive data curve) in the pulsar signals. He reasoned that if he got a Doppler-shifted amplitude variation that followed a pattern, almost certainly it would mean that the perturbations causing the variable arrival times were due to planets orbiting the pulsar.

But before he proceeded, Wolszczan knew that he needed someone to help him get better position coordinates for the pulsar. For those coordinates, he turned to Dale Frail at the National Radio Astronomy Observatory's Very Large Array (VLA), a network of 27 radio telescopes that travel along three arms of Y-shaped railroad track in New Mexico. Back in his office at Cornell one September afternoon in 1991, with the data from Frail in hand, Wolszczan recognized that there had to be two, possibly even three, planets orbiting PSR 1257 +12. From analyzing and reanalyzing the periodicities in the data, Wolszczan realized that the masses of

The National Radio Astronomy Observatory's Very Large Array (VLA) west of Socorro, New Mexico, is made up of 27 movable, 25-meter radio dishes that can be positioned in a configuration of up to 36 kilometers across. The dishes are moved along three railroad arms, which make up the array's Y-shaped pattern. (Photo courtesy M. Colleen Gino.)

the objects causing the perturbations and their clockworklike regularity suggested objects the size of Earth in orbit around this unique pulsar. (He estimated the largest to be only five times the mass of Earth.)

Around the same time in 1991, Andrew Lyne laid claim to similar perturbations around PSR 1829 –10, a pulsar near the galactic center. Lyne was set to present his data, collected at the Jodrell Bank Observatory, at a meeting of the American Astronomical Society (AAS) in Atlanta in early 1992. Wolszczan was also scheduled to speak at the meeting. The media were already circling, waiting for the announcement of the first planet found outside the Solar System. But when Lyne took the podium, he stunned the crowd by announcing that his "planet" was nonexistent. Due to a basic error, he reported, his team had failed to account for Earth's own movement around the Sun, so the only perturbations found in the pulses from PSR 1829 –10 were simply the result of miscalculations here on Earth. "In 1829 –10," says Lyne, "we corrected for 99.9 percent of Earth's movement. It was the remaining .01 percent that we failed to do correctly." [9]

Surprisingly, the attendees at the AAS meeting awarded Lyne with a standing ovation, in large part because he had been so gracious and forthright in recanting his "discovery." As the applause for Lyne subsided, Wolszczan took the podium to present his data on the perturbations and possible planets around PSR 1257 +12. Fred Rasio, a theoretical astrophysicist at M.I.T., was at the meeting and remembers that Wolszczan outlined his findings with details about all the checks and tests that he had done before reaching his first unconfirmed conclusions. "The audience's response was that they thought at least that guy is being careful," says Rasio, but "there was a lot more emotion to Andrew Lyne's talk." [10]

After three more years of observations at Arecibo, Wolszczan was able to offer irrefutable proof that three planets indeed circle PSR 1257 +12; the proof was demonstrated by gravitational interaction of the planets with each other, a gravitational perturbation that Rasio himself had predicted. At the 1999 Bonn conference, Wolszczan presented the latest statistics on the three planets, based on new observations made at Arecibo as recently as the late 1990s. All three planets have circular orbits, and all are coplanar (meaning they orbit the pulsar in the same plane of the sky so that if being viewed edge-on, they would eventually occult or eclipse). The innermost planet was the hardest to detect, Wolszczan explained, because its small mass meant that it would create the smallest perturbations. The innermost planet's Moon-like mass orbits the pulsar every 25.3 days. The middle planet is the equivalent of five Earth masses and has an orbital period of 66 days. And the outermost planet is the equivalent of four Earth masses and has an orbital period of 98.2 days. None are candidates for colonization.

Wolszczan also reported that, based on the dynamics of the system, it is clear that the planets were formed from supernova remnants in orbit around the pulsar in much the same way planets form in circumstellar disks around stars like our own. He believes that because these three planets circle a pulsar, instead of a normal hydrogen-burning star, there is an astro-political bias against their assignment as planets. "I don't understand it," says Wolszczan. "Planets are planets dynamically, even if they are flying rocks around a pulsar. For all practical purposes they are planets. If you change the definition, and you say that anything that orbits a star that is not a Sun-like star is not a planet, then they are right. But it's a matter of definition."

To date, the PSR 1257 +12 planets are the only known system of terrestrial-class planets outside our Solar System. There is, however, at least one other confirmed planet orbiting a pulsar. But its mass indicates that it is at least five times the mass of Jupiter, and was most likely scavenged by the pulsar from a nearby star. In 1993, a year after Wolszczan's announcement, Donald Backer, a radio astronomer at the University of California at Berkeley, reported findings based on observations of PSR 1620 −26, using the National Radio Astronomy Observatory's 42-meter radio telescope in Green Bank, West Virginia. Using data accumulated on PSR 1620 −26 since 1988, Backer found what, in timing, should have been an ordinary millisecond pulsar being orbited every 191 days by a white dwarf companion, roughly a third the size of our Sun. But a signal perturbation indicated that the two-body system was also being orbited by a third body. At the time, Backer reported that this third body was in roughly a 100-year orbit around both the white dwarf and pulsar, and had an estimated mass of at least ten times that of Jupiter. (Its mass has subsequently been determined to be about half that—5 Jupiter masses). It is located in M4, a globular cluster some 1,000 parsecs away in the constellation of Scorpio. Globular clusters are very old, dense concentrations of stars. In such concentrations, it is theorized that it would not be unlikely for a pulsar to interact with either a normal solar-type star being orbited by a normal planet, or a white dwarf being orbited by a normal planet. The pulsar would join the party as an uninvited guest, gravitationally binding both its new companion star and its planetary companion to its own sphere of influence.

WHITE DWARF

A white dwarf is the exposed core of a low-mass star that has essentially collapsed onto itself after nuclear fusion has finished. Due to a combination of trapped heat left over from earlier nuclear fusion and gravitational contraction of the stellar core itself, a white dwarf is formed at temperatures over 10,000 degrees Kelvin. This causes its white-hot appearance. A white dwarf above 1.44 solar masses would become even more dense, and as explained earlier, contract to form either a neutron star or a black hole. It is believed that our local solar neighborhood is filled with white dwarfs that are simply too dim to spot. In the course of 10 billion years, a white dwarf will lose all its luminosity and lapse into an eternal state as a black dwarf (or the cooled collapsed core of a star that no longer emits light). Unless a black dwarf collides with another star, its physical integrity as a solid mass will theoretically remain intact for an infinite period of time.

For his part, Andrew Lyne remains convinced there is something bizarre in the timing of PSR 1829 −10, but attributes the anomaly to a basic asymmetry in the shape of the pulsar itself, which could be causing it to precess (wobble in the manner of a spinning top) as it rotates. And though he's certainly not disheartened about his misadventures in planet hunting, he admits he would like to find a pulsar planet eventually. Most pulsar astronomers, on the other hand, couldn't care less about planets orbiting pulsars; according to Bell Burnell, they are interested only in planets to the extent that they affect the dying star itself.

Rasio points out that "it's absolutely clear that something like the Wolszczan system is very rare. Many other millisecond pulsars have been searched to a level of timing accuracy such that anything like the Wolszczan system would have been detected. But they are clearly not there. It obviously takes something relatively special in the evolution of the pulsar for planet formation to occur. What it is we don't know."

Abell 39, lying well over 2,100 parsecs away in Hercules, is the 39th entry in a catalog of large nebulae discovered by American astronomer George Abell in 1966. It is unique as a planetary nebula for its rare, spherical symmetry and the fact that its white dwarf is visible directly in the center of the nebula. Earth's closest known white dwarf, however, orbits Sirius, the brightest star in the sky, lying only 2.6 parsecs away in the constellation Canis Major. (WIYN/NOAO/NSF.)

At the time of this writing, well over 60 planets are known to circle distant stars, but only 4 orbit pulsars. The rest orbit stars pretty much like our own. Extra-solar planets are now being discovered at a rate of about 2 per month, meaning that by the time this book is in your hands, there will likely be more than 70 known extra-solar planets. It's only a matter of time before astronomers find Earth-like planets circling stars like our Sun.

Life on them is a long shot. But even so, with an estimated 200 billion stars in our galaxy alone, the odds are hugely in life's favor. The idea that out of a Universe at least 10 to 13 billion light years across, we are the only cognizant, communicative entities is very nearly as inconceivable as the size and scope of spacetime itself. The question facing this generation and the next, however, is not whether there is intelligent life in the Universe, but rather whether there are Earth-like planets circling other Sun-like stars. Without such extra-solar terrestrial-type planets within habitable orbits around Sun-like stars, the widely envisaged plethora of extraterrestrials voyaging across the cosmos will forever remain science fiction.

Within the next 50 years, astronomers should literally resolve this fundamental question, not through the off-chance detection of an extraterrestrial intelligent laser or radio beacon, but via real-time direct observation of Earth-like planets circling nearby stars. At present, the only known extra-solar terrestrial-mass planets orbit a lifeless pulsar, which in and of itself hardly constitutes a habitable zone. Whether astronomers will ultimately find Earth-like planets in habitable orbits around other Sun-like stars remains very much an open question. However, Bell Burnell, now a visiting professor at Princeton University, remains encouraged: "There are several very good reasons why there should not be planets around a pulsar, and yet there are," she says. "That suggests that planets can be made quite late in the life of a star. To me that says that planet formation is relatively easy. If it can happen post-supernova, planet formation is not too difficult a mechanism and must operate quite well. Planets are ubiquitous."

1   Peterson, Charles J. 1997. "Magellanic Clouds." In *History of Astronomy: An Encyclopedia.* John Lankford (ed.). New York: Garland Publishing, Inc.: 317.

2   Reid, Mark, astronomer, Harvard Smithsonian Center for Astrophysics. Interviewed on January 19, 2000.

3   North, John 1994. *The Fontana History of Astronomy & Cosmology.* London: Fontana Press: 545.

4   North, John 1994. *The Fontana History of Astronomy & Cosmology.* London: Fontana Press: 566.

5   Hughes, John, astrophysicist, Rutgers University at Piscataway, New Jersey. Interviewed on January 14, 2000.

6   Bell Burnell, Jocelyn, astronomer, Open University, Milton Keynes, U.K. Interviewed on September 2, 1999, at the Pulsar 2000 Colloquium, Bonn, Germany.

Bell Burnell, Jocelyn. Public lecture at *Deutsches Museum*, Bonn, Germany, September 2, 1999.

7   Hewish, Antony, Cavendish Laboratory physicist, Cambridge University, U.K. Interviewed on September 1, 1999, at Bonn, Germany.

8   Wolszczan, Alexander, astronomer, Penn State University. Interviewed on August 31, 1999, at Bonn, Germany.

9   Lyne, Andrew, astronomer, Jodrell Bank Observatory/University of Manchester. Interviewed on September 2, 1999, at Bonn, Germany.

10   Rasio, Fred, theoretical physicist at M.I.T. Interviewed on March 7, 2001.

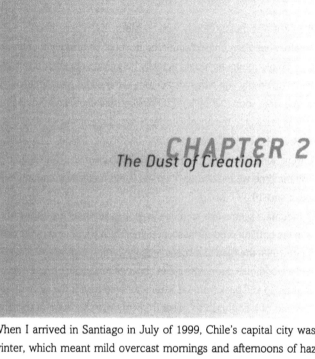

CHAPTER 2
The Dust of Creation

When I arrived in Santiago in July of 1999, Chile's capital city was in the midst of a normal winter, which meant mild overcast mornings and afternoons of hazy sun. For a country that prides itself on having some of the best astronomical seeing in the world, it looked more like Geneva in December, completely socked in and gray. But an hour's flight north of Santiago, the weather dramatically improves—or so I was told—and I was soon touching down in La Serena, a small resort city on the coast. Just inland, and within easy driving distance of La Serena, lie two of the world's best observatories: the Cerro Tololo Inter-American Observatory (CTIO) and the European Southern Observatory's facility at La Silla.

My first stop was CTIO. Within hours of arriving in La Serena, Tito, my taxi driver, and I were making our way out of a cacophony of honking horns, celebrating Chile's victory over Colombia in that afternoon's soccer match, and into a sun setting over the fertile green Elqui Valley. Surrounded by chirimoya fruit, apricots, grapes, and papaya trees, people dotted the roadside, waiting for buses to ferry them home from their Sunday outings. Then, the verdant valley turned drier and the road wound higher, until we were almost an hour out of town. Finally, Tito turned onto an unmarked road leading to the territory owned by the CTIO and into the barren foothills of the Andes, some 85 kilometers southeast of La Serena. Two men on bicycles were being trailed by a tired-looking German shepherd a couple of kilometers shy of the guardhouse entrance to CTIO's private, seemingly never-ending 35-kilometer stretch of lonely rutted road. Totally devoid of human habitation, its only signs of life were cacti and, every 10 kilometers, emergency radiophones. It's certainly no place in which to speed, as there are hairpin turns, steep drop-offs, and no guardrails. (Two British journalists returning from an observatory visit around midnight one evening found themselves at the bottom of a ravine after failing to navigate a curve.)

On this Sunday in mid-July, it was mostly cloudy with no Moon. Even here, in one of the driest places on Earth, the weather has been unpredictable, so it would be unlikely that any of the nine telescopes atop Cerro Tololo mountain would actually be making observations. Even so, when we arrived, Tito, who'd obviously come this way before, dimmed his headlights out of courtesy to the astronomers. We were soon met by a CTIO pickup truck that escorted us up a steep two-lane blacktop driveway leading to the observatories. From there, it was off to CTIO's *cantina* and *dormitorio*, archetypes of the mundane in a surreal landscape. Inside, Tito greeted the kitchen staff as if they were long-lost family; then we wolfed down a quick dinner of *empanadas* and vegetables. By the time we headed back up the 2,200-meter high mountain to the Blanco 4-meter telescope, it was 10:00 P.M.

In its barren state, the mountain seems like an unlikely spot to serve as a window onto supernovae, stellar nurseries in the birthing process, nascent planetary systems, and other wonders of our galaxy. But until late 1998, the Blanco 4-meter was the largest optical telescope in the Southern Hemisphere, and a workhorse during the initial observations of Supernova 1987A. In fact, Tito had hauled journalists up and back from La Serena for weeks after the story of that discovery broke. But now, instead of being used to observe overly massive stars in the final

Star trails are evident in this long-exposure photograph of four domes of the Cerro Tololo Inter-American Observatory, seen here under a particularly clear sky from atop Chile's Cerro Tololo mountain. (Photo courtesy CTIO.)

death throes of a supernova, the Blanco was being used to view stars much like our own. These stars are much younger planet-forming versions of our own Sun, still in the process of writing another chapter in our galaxy's never-ending cycle of death and creation.

Wherever the sky looms large—Australia, the American West, Central Asia, South Africa, Florida, Hawaii, much of the Mediterranean, the Middle East, and certainly here in Chile—interest in the celestial sphere seems almost innate. Some 150 years ago, before the advent of modern industry and the automobile, Earth's skies were cleaner, clearer; and, as a result, between dusk and dawn, the stars seemed to shine brighter than they do today. In *The Adventures of Huckleberry Finn*, Mark Twain describes how Huck and Jim while away the evenings floating down the Mississippi River on a homemade raft. They spend much time marveling at the nighttime sky. It is estimated that the astronomical "seeing" in the lower Mississippi River valley at that time was 50 to 75 percent better than today. In the book, Huck posed the same questions we ask today: "We had the sky up there, all speckled with stars, and we used to lay on our backs and look up at them, and discuss about whether they was made, or only just happened—Jim he allowed they was made, but I allowed they happened; I judged it would have took too long to make so many. Jim said the moon could a laid them; well, that looked kind of reasonable, so I didn't say anything against it, because I've seen a frog lay most as many. . . ."

But even here in Chile, more than 2,000 meters above sea level, there are nights when the seeing is less than spectacular. So upon entering the control room of the Blanco 4-meter telescope, Tito and I did not interrupt astronomers at their work, but rather an in-progress game of sunflower-seed poker being played by an observing team from the University of Florida who were waiting for the cloud cover to clear. With a CD player blasting a seventies-era party tune, it was hardly the stale, humorless environment often presented when Hollywood goes to the observatories. Scott Fisher and James De Buizer, at the time of our 1999 visit both graduate students, were there in support of other astronomers who were using OSCIR, a high-resolution infrared imager that they attach under the 4-meter telescope each night. Built by principal investigator Charles Telesco and the OSCIR team at Florida, OSCIR, an acronym for Observatory Spectrometer and Camera for the Infrared, has been attached to its share of high-powered telescopes in the recent past, including the Keck Observatory in Hawaii, which at some 4,200 meters altitude does some of the best infrared "seeing" in the world. "Hawaii is plus 19 latitude, and [the Keck] can barely see down to minus 50," says Fisher. "So here at Blanco, we have the whole extra 40 degrees of sky that nobody [else] can see right now."[1]

Yet because the Blanco 4-meter telescope has "diffraction-limited" optics, optics that exceed its own theoretical capability to resolve a point in space, OSCIR is also able to perform at its optimal level, provided the right local "seeing" conditions are available on any given night. But at CTIO, objects in the Southern Hemisphere can be viewed through less of Earth's atmosphere than in Hawaii, giving Blanco more of a straight-up shot and less atmospheric disturbance to interfere with the quality of its images. So even though Keck offers a wonderful site

Inside the dome of the Blanco 4-meter telescope CTIO Observatory. It began operation in 1976 and is named for American astronomer Victor Manuel Blanco, the observatory's first director. The telescope was designed as a southern counterpart to the 4-meter telescope at Arizona's Kitt Peak National Observatory. (Photo courtesy CTIO.)

for infrared astronomy, Blanco offers OSCIR both atmospheric and optical quality that pushes the observers and their infrared imager to the limit of their abilities.

PLANETARY BIRTH

When they can, Fisher and De Buizer also use OSCIR to observe the planet-formation processes that are taking place around several nearby stars. This raises a question: Though we know that pulsar planets most likely form from the debris left over in the wake of a supernova explosion, how did our own and other planetary systems form?

The answer requires a look back to the Sun's primordial beginnings. Some 5 billion years ago, our Sun and its resulting planetary disk collapsed from a dense molecular cloud of hydrogen and helium. The initial collapse of this cloud of gas and dust, which lasted an estimated 100,000 years, was likely triggered by a nearby—relatively speaking—supernova explosion. As the Sun and its disk collapsed, it gained angular momentum, or rotation, and kinetic energy, and took on the characteristics of a young circumstellar disk. Then, as our budding Solar System's molecular cloud contracted, it heated up. Dust grains embedded in the cloud caused it to contract even further, as the grains emitted infrared radiation that cooled the cloud and thus allowed it to continue to contract.

The Collapse of the Sun

Within 100 million years of its initial contraction, the resulting burgeoning stellar core began thermonuclear fusion. This was followed, an estimated 4.6 billion years ago, by this "protosun" fully switching on to its main-sequence phase, fusing hydrogen into helium. That fusion continues today.

If we were looking onto our Solar System from above, it would appear that the whole thing, including the Sun, was moving counterclockwise. This movement was initially caused by a physical law known as the *conservation of angular momentum*. The law describes the manner in which the gravitational collapse of the molecular clouds triggered the beginnings of our Solar System, and should also apply universally for planetary formation around other Sun-like stars. After the collapse of large dark clouds of molecular hydrogen and dust, our evolving Solar System was beset by an infall of dust and gas toward a central point—our nascent protostar. This infall is affected by physical laws to conserve energy in the same way that a figure skater, going into a spin, speeds up as she draws in her arms. The result was that the material that formed our early Solar System began to flatten, or "pancake," around a central axis of rotation. This central point of rotation remains our Sun; but in the early stages of the Solar System's formation, this rotational movement triggered a continual gravitational collapse of dust and gas that eventually formed a proto-sun and a nascent planetary disk that orbited it. It was from this disk that our nine planets, their more than 70 known satellites, and other detritus left over from our system's construction (such as asteroids, comets and meteors) were formed. Most of our Solar System's mass is tied up in our own star (that is, our Sun). Jupiter, the gas giant, is 2.5 times more massive than the rest of the planets combined. It and the other three gaseous planets—Saturn, Neptune, and Uranus—likely formed within 10 to 20 million years after the Sun started its hydrogen-burning phase. Earth was something of a laggard, forming some 100 to 200 million years after these gas giants, or almost 4.5 billion years ago.

According to a theory confirmed in 1998, the planetary formation process starts when dust grains—like those visible with instruments such as OSCIR—begin colliding and accreting via electrostatic attraction. That year, a German-designed microgravity dust experiment, conducted aboard the U.S. space shuttle Discovery, proved that the origin of our own planetary system stemmed from the collision of micron-sized particles of silicon dioxide (SiO_2)—essentially tiny grains of glass. At room temperature, these particles begin colliding at only millimeters per second, and stick together through what is known as the *van der Waals force*, an electrostatic force that allows for the mutual attraction of individual molecules.

VAN DER WAALS FORCE

Dutch physicist Johannes Diderik van der Waals, who won the 1910 Nobel Prize in Physics for his work on the various states of matter in its liquid and gaseous forms, also studied the interactive forces that influence individual atomic molecules. Known as dipole moments, for their very weak polarizing and electromagnetic effect on each other's individual electrons, the van der Waals force is known to cause molecules to repel or attract each other when in close proximity. This is quite a different interaction from the interactions that take place due to valence bonds on the atomic level, or bonds that come about when one electron circling an atom's nucleus takes part in chemical bonding with another atom. A common effect of the van der Waals force can be seen in the manner in which the molecules of water will appear to have an inexplicable attraction when riding up on a full drinking glass, appearing to form an unexplainable bulge out beyond the lip of the glass.

The German Space Agency's Cosmic Dust Aggregation (CODAG) experiment consisted of dust grains interacting at room temperature in a low-pressure, vacuum-sealed, 2-liter box, which was loaded into the shuttle's payload bay. Within minutes of activating the experiment, a stereoscopic microscope linked to the box recorded helical-shaped—that is, spiral-shaped—dust particles colliding and forming asymmetrical shapes. Jürgen Blum, CODAG's team leader, says that, in theory, within one year, these dust particles could grow from microns to millimeters, and within 100,000 years, from millimeters to kilometer-sized planetesimals. Then mutual gravity would take over.[2]

PLANETESIMALS

Planetesimals are solid particles of the Solar System's primordial debris, and can range in size from a few microns to 100 kilometers in diameter. Most meteoroids (fragments of comets and asteroids) and some asteroids (which can be small, but often are the size of minor planets) would fall into the category of planetesimal. Planetesimals above a kilometer in diameter are subject to the forces of gravity. Within a few million years, asteroidlike bodies from 100 to 1,000 kilometers in diameter form and begin what is termed a "runaway accretion process," whereby huge chunks of matter start colliding. (Imagine drunk drivers circling a bumper car rink on a 600-million-year ride.) This heavy bombardment period results in the formation of larger terrestrial-sized chunks, like our Earth and the other inner planets.

Within a few million years of its formation, our inner Solar System likely consisted of several hundred bodies, each a 1,000 kilometers in diameter. Although the velocity of the planetesimals increases almost exponentially as they continue to gain mass, an inner planetary system like ours would be slower to assimilate massive planetesimals into fully formed planets. As a result, scientists estimate that it took as long as 200 million years from the onset of microscopic agglomeration of dust in our own solar nebula before the Earth emerged in its primordial shape.

WRITING PLANETARY HISTORY IN DUST

Until the late 1990s, there had been something of a disconnect between the planetary science necessary to understand our own Solar System and the astronomy and technology used to observe planet formation that's taking place around nearby stars. Planetary scientists are still trying to sort out the vagaries of the Solar System. They have just begun to learn the formation mechanisms behind it, marked by the launch of a decades-long program of exploring the outer planets of our Solar System, which began in the late 1970s with NASA's early planetary flyby missions.

To date, progress in reaching a fuller understanding of our Solar System has been slow. Not surprisingly, then, in 2000, scientists in the U.S. were dismayed and disheartened to learn NASA had reneged on funding for a flyby mission of Pluto and its moon Charon. Consequently, while planetary scientists are excited by the rash of recent discoveries of planetary and debris disks around other stars, they also feel frustrated by the lack of funding for remote exploration of what has come to be thought of as our own backyard. They argue that we must first nail down the specifics of the evolution of our own Solar System with some degree of certainty. Doing so would result in a much better benchmark against which to characterize budding planetary disks around stars that are both similar and sometimes very dissimilar to our own. After all, it has only been six years since the first indirect detection of a planet circling another Sun-like star. And most of what is known about extra-solar planet formation began to manifest itself—unexpectedly—18 years ago, when the Infrared Astronomical Satellite (IRAS), a joint U.S.-Dutch-British satellite launched in early 1983, made a serendipitous find.

The Vega View

One of the primary goals of the IRAS mission was to survey bright main-sequence stars similar to our own. The satellite found that Vega, a young (350-million-year-old) star two to three times the mass of our Sun, emitted 1,000 times more radiation than expected. The source of this infrared radiation seemed to be heated dust in what is now thought to be an emerging planetary system. In other words, the dust around Vega is due to accretion and to the resulting debris from colliding planetesimals.

More than 100 stars have been observed to have Vega-like dust around them. Zodiacal dust left over from the original collapse of the stellar nebula, or first-generation dust, in very young planetary systems (less than 10 million years old) usually becomes dissipated by radiation and stellar wind emitted from their budding parent star. Yet oft-repeated observations of Vega seem to indicate that it rid itself of its first-generation primordial dust long ago and moved

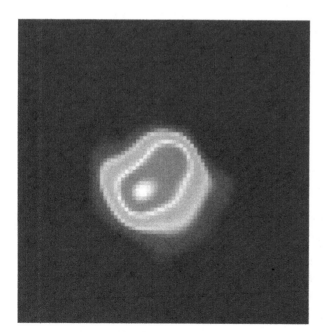

This submillimeter image of the bright star Vega (Alpha Lyrae) and its extended protoplanetary disk was taken by the Submillimeter Common User Bolometer Array (SCUBA) camera at the 15-meter James Clerk Maxwell Telescope (JCMT) in Hawaii. (James Clerk Maxwell Telescope owned by the U.K., Canada, and the Netherlands.)

on to the latter stages of planet formation. (It is known that debris disks older than 100 million years are a telltale sign of planet formation.) Without debris and second-generation zodiacal dust formed from colliding planetesimals (like that found in our own Solar System and around Vega), there would be no debris found around stars more than 100 million years old.

The Beta Picture

The disk of dust surrounding Beta Pictoris, 19.28 parsecs from Earth, is more readily apparent than that of Vega. Lying in the southern constellation of Pictor, the disk around this high-mass star is among the most studied of all the newly formed planetary disks. In 1984, its disk was found to be edge-on; three years later, it was the first planetary disk to be imaged in the infrared. And in the late 1990s, images from the Hubble Space Telescope revealed that one side of the disk is 20 percent more extended than the other. Some theorists believe the most likely explanation for this "bulge" is that a passing star zipped by some 100,000 years ago and caused the distortion. Astronomers are looking for the culprit from among 186 candidates in its general vicinity. (By measuring any given star's current trajectory across the sky, astronomers can usually extrapolate the path that is has taken over periods ranging over hundreds, even thousands, of years.) One thing for sure is that Beta Pic, as it is affectionately called, is a lone wolf. At least 100 parsecs from the nearest star-forming region, this 100-million-year-old star has a disk that may harbor at least one planet and a multitude of comets that stretch out to 1,400 astronomical units (AUs).

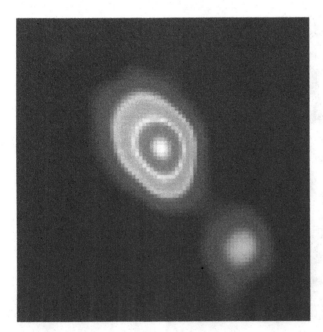

A submillimeter image of Beta Pictoris, also taken by the SCUBA camera at JCMT, clearly shows this southern star's protoplanetary disk, in which some astronomers believe a companion planet ten times the mass of Earth has already formed. (James Clerk Maxwell Telescope owned by the U.K., Canada, and the Netherlands.)

ASTRONOMICAL UNIT

The average distance between here and the Sun is 1 AU, about 93 million miles. By contrast, the orbital boundaries of the nine planets of our Solar System extend to almost 40 AU. Beta Pic has a central hole in its disk at 35 AU, almost the mean distance at which Pluto orbits the Sun.

"We don't know whether disks like Beta Pic are examples of stars forming planetary systems or of failed planetary systems," says Fisher, who noted that OSCIR had also imaged Beta Pic. "We don't know whether these disks form planets all the time or none of the time. We believe that the dust particles, of smaller silicate dust, are what accreted into planets. It would be very hard to have planets without it." Yet we now know that, 4.5 billion years after the formation of our own Solar System, we are still surrounded by tiny particles of zodiacal dust, which are caused primarily by the collisions of asteroids in the asteroid belt between Mars and Jupiter. Astronomers claim that this dust still manifests itself, however weakly, just after dusk in the reddish hue of our vanishing Sun.

Dust, then, is the key to both planet formation and the collapse of protostars around which planets can form. Analysis of data collected in the late 1990s by the ESA's Infrared Space Observatory (ISO), which was deployed in Earth orbit in 1995, now shows that protostars in the midst of opaque layers of dust may ignite sooner than had been previously believed. The ISO data clearly indicate that VLA 1/2, a newly forming star in the Orion Nebula, some 370 parsecs away, has already ignited. This newborn star is surrounded by a dusty disk that extends to 5 AU, roughly the distance at which Jupiter orbits our own Sun. Here in this surrounding protoplanetary "placental envelope," ESA astronomers say, it is likely that planets will form. (According to an ISO sample of 84 nearby main-sequence, hydrogen-burning stars, 60 percent of those under 400 million years old have debris disks of the sort required to make planets.)

Scott Fisher in the ground floor control room of the Blanco 4-meter telescope. (Photo by Bruce Dorminey.)

While the ESA astronomers estimate that this surrounding area must have average temperatures of at least 500 degrees Kelvin, their data also reveal cooler, ice-forming regions that contain dust grains shrouded in frozen mixtures of water, methane, carbon dioxide, and possibly even methanol.

Following a Dusty Trail

Dust disturbances in a planetary disk can also be used to trace newly formed planets. A team led by Nick Gorkavyi, an astrophysicist at the NASA Goddard Space Flight Center in Maryland, has come up with a new method to identify planets that are plying their way through the dust of young planetary systems. Given the long life of such dust clouds surrounding these systems, Gorkavyi believes that, thanks to today's technology, it is possible to identify a planet's mass and orbital characteristics simply by reading the patterns of swirls, arcs, and clumps it leaves in its wake. He surmises that, rather than a perturbation from a passing star, the asymmetrical warping observed in the dust surrounding Beta Pic is most likely caused by a planet at least ten times the mass of Earth circling the star in a very long asymmetrical orbit. Gorkavyi and his team also believe that Vega may have a planet twice the size of Jupiter circling the young star, lying in an equally long orbit, farther out than Pluto circles the Sun.[3]

Gorkavyi plans on using the "dust-tracking" method to look for evidence of planets in orbit around young stars that may lie as far as 30 parsecs away. By observing how the asymmetrical dust structures kicked up by the planets also revolves around the young stars, the NASA team hypothesizes that in several years they will be able to more accurately confirm the planets' orbital parameters and masses.

Where the Trail Begins

How these planetary dust particles form in the first place has always been a conundrum for theorists. Just as a supernova may have caused the collapse of the molecular cloud and led to our Sun's own formation, our Solar System's formation from a protoplanetary disk may have been speeded along by a nearby gamma-ray burst (GRB).

GAMMA-RAY BURSTS

GRBs were discovered over 30 years ago during the height of the Cold War, after the U.S. had launched a series of VELA satellites in very high elliptical orbits around the Earth. The VELA program, funded by the U.S. Department of Defense and the U.S. Atomic Energy Commission, spanned the decade from 1969 to 1979. The program's objective was to search for clandestine nuclear detonations on Earth and from space, and their resulting X-ray and gamma-ray emissions. During this era, it was thought that the Soviets might try to hide a nuclear test on the far side of the Moon. What the satellites observed instead were sporadic flashes, or bursts, of gamma rays from deep space.

The astronomical community has been scratching its head ever since the Pentagon revealed this finding. Gamma rays, the highest form of electromagnetic radiation, can easily go through several inches of steel, yet are normally absorbed in Earth's atmosphere and are subsequently rendered harmless to humans. In contrast, GRBs, which last from milliseconds to minutes, produce more energy than our Sun will in the whole of its 11-billion-year life. And GRBs are ubiquitous; during more than a decade of service orbiting Earth, NASA's Compton Gamma Ray Observatory cataloged more than 2,500 GRBs. Gamma-ray bursts remain the most massive known explosions in the Universe. Some 150 extant theories describe what causes GRBs, but two are put forth most often. One argues that GRBs stem from colliding neutron stars, the dense remains of stellar cores left over from supernovae explosions. Neutron stars frequently coalesce, and eventually merge, after millions of years circling each other in a double neutron star binary system. This merger may be one likely source of GRBs. The other source could come from energy released via a hypernova, or an extremely violent and unusually energetic form of a supernova. Hypernovae are thought to result from the collapse of a supermassive star in very dense star-forming regions. The cores of hypernovae may in fact collapse into black holes, and in the process, release GRBs.

Heating Our Solar Nebula

Brian McBreen, an astrophysicist at University College in Dublin, Ireland, is more interested in the effects of GRBs than their cause.[4] He believes that the Solar System's nascent chondrules were made molten in a matter of seconds after a nearby GRB (within 100 parsecs) blasted our young Sun's circumstellar disk. He concedes that the probability of such a GRB occurring within 500 parsecs is only 1 in 1,000.

> CHONDRULES
> Chondrules, essentially millimeter-sized glass spheres, contain volatile silicate minerals, such as olivine (a group of rock-forming greenish-colored silicates made up of magnesium and/or iron) and pyroxene (also a group of rock-forming silicate minerals that usually form igneous rocks). Pyroxenes can also contain combinations of minerals such as calcium, sodium, magnesium, iron, and/or aluminum. Such chondrules, with their complex silicate chemistries, are the building blocks of both planets and asteroids. They are often found embedded in meteoric stones.

McBreen asserts that if his theory is correct, chondrules should retain a chemical GRB signature that shows that they were made in large quantities, and then crystallized rapidly from molten or partially molten drops. Chondrite meteorites, named for the chondrules that are often found in their makeup, frequently lie in the asteroid belt between Mars and Jupiter. But even now, such meteorites are sometimes kicked into the inner Solar System and the path of Earth. Almost 90 percent of the meteorites that survive entry through Earth's atmosphere until impact are chondrites.

Contrast McBreen's theory with the conventional scenario describing our Solar System's formation, which relies on electromagnetic shock waves and lightning to heat the dust particles in sufficient amounts to melt them into chondrules. As planetary scientists have long acknowledged, there is a weak link in such traditional ideas about the formation of chondrules. Specifically, grains of dust would have to be heated to temperatures of 1,500 degrees Kelvin in order to melt en masse. Lightning or electromagnetic shock waves alone can't account for such a process. McBreen points out that if this GRB-chondrule formation scenario is correct, based on the frequency of GRBs in our "neighborhood," then only 1 planetary system in 1,000 should have evolved like our Solar System. That would certainly constrain planet formation.

WHERE WE GO FROM HERE

Clues as to whether our Solar System is unique, or merely one of many quite similar to it, may come from an unlikely setting in one of the most inhospitable places on Earth. Several hundred kilometers north of La Silla and CTIO, in the far end of Chile's Atacama desert, is the site that eventually will host an interlinked $560-million 64-element array of 12-meter parabolic antennas. The Atacama Large Millimeter Array (ALMA), backed equally by the U.S., Europe, and Japan, will enable astronomers to understand the earliest phases of star and planet formation in submillimeter-wavelength detail. Scheduled to be fully operational by 2010, it should be capable of detecting giant gas planets that are literally in the process of formation.

Astronomers, however, are not waiting for ALMA. NASA's Space Infrared Telescope Facility (SIRTF), due to be launched into orbit in July 2002, will begin collecting data on possible circumstellar disks surrounding hundreds of stars, many of which will be similar to our own. SIRTF's observation of these stars will comprise 350 hours of the telescope's five-year mission. Michael Meyer, an astronomer at the University of Arizona in Tucson, leads this disk-hunting aspect of the SIRTF mission. Meyer notes that with few exceptions, SIRTF would not be sensitive enough to detect zodiacal dust around older stars like our own. But it should finally enable theorists to bridge the gap in planetary disk evolution, because SIRTF will allow for observation of stars between the earliest phases of planet formation and what Meyer termed second-generation debris disks, which approach analogs of our own Solar System. When his team begins data analysis in 2003, Meyer hopes that their work with SIRTF will finally enable planetary scientists to place the Solar System in context.[5]

By the end of this decade, planetary scientists and theorists, who have eagerly awaited each new announcement of an extra-solar planetary disk or planet, may finally be able to accumulate a pertinent critical mass of data on planetary formation outside the Solar System. It is hoped that such data will enable them to fill in the blank page of planet formation history, and thus be able to determine how our Solar System fits in with those around other stars.

An artist's impression of the Atacama Large Millimeter Array, a collection of up to 64 12-meter antennas to be located on a high dry plateau at the northern end of Chile's Atacama desert, at an altitude of 5,000 meters. Fully operational in 2010, it will be used to image planet formation in its earliest stages. (European Southern Observatory.)

Meanwhile, OSCIR and other instruments continue their surveys. OSCIR's biggest find to date came in March 1998, when Ray Jayawardhana, formerly an astronomer at the Harvard Smithsonian Center for Astrophysics, used the Blanco 4-meter telescope and OSCIR to produce an image of a room-temperature disk of dust around the young (10-million-year-old) star HR 4796A. Several times more massive than our Sun, HR 4796A is located in a binary star system in the Centaurus constellation, some 67.1 parsecs away. Like Beta Pic, the star is surrounded by what has been called a spectacular, nearly edge-on circumstellar disk, highly visible at the upper end of the mid-infrared region. But recent Hubble Space Telescope (HST) images of HR 4796A show that it also has a peculiar asymmetry that causes one side of the disk to glow brighter than the other. The University of Florida team at CTIO concludes that this could be the result of a recently formed planet shifting the dust-disk slightly off center.

Scott Kenyon and Kenny Wood at the Harvard Smithsonian Center for Astrophysics did a computer analysis of photos taken of HR 4796A by the HST. They have concluded that it has a ring of debris circling in its outer disk, just as our system has remnants of such debris today. A similar debris field in orbit around HR 4796A may signal that our own system's formation is being mimicked continually across our galaxy. It also shows that planet formation is possible in binary star systems, an important point to consider given that an estimated 60 to 80 percent of all stars are binaries (or double stars).

The Kuiper Belt

Our own system's "ring of debris" is known as the Kuiper Belt, named for Gerard Kuiper, who in 1951 theorized that such a belt would likely contain a plethora of planetesimals and small asteroid-like bodies that would circle well beyond the orbit of Pluto. In fact, rather than a *bona fide* planet, Pluto is thought by some to be a large planetesimal that strayed in from this belt of debris left over from our own planet-formation process. As of this writing, Pluto remains classified as a planet, but planetary scientists continue to argue whether it should be downgraded to planetesimal.

> ### PLUTO IN PERSPECTIVE
> *We often forget how little we know about our own Solar System. It was only a little more than 70 years ago that Clyde Tombaugh, an amateur astronomer at Lowell Observatory in Flagstaff, Arizona, discovered Pluto, the frozen ball of ice and rock in our system's outermost fringes. Charon, its only known moon, was found in 1978.*
>
> *Tombaugh's triumph had been put in motion by the dedicated efforts of another amateur astronomer, Percival Lowell, who was so determined that he had set up his own observatory on a hill overlooking what was then the whistle-stop northern Arizona town of Flagstaff. Lowell was intent on finding Planet X, as the ninth planet from the Sun had been dubbed, but unfortunately died in 1916 after over ten years of fruitless searching. After Lowell's death, his observatory languished under legal and financial troubles for several years, until Tombaugh, a young amateur astronomer and photographer recruited from Kansas, began searching anew in the spring of 1929. He succeeded in finding the planet in January 1930.*

Pluto's name is derived from two sources: the initials of Percival Lowell, and the suggestion of a British schoolgirl who reasoned that by virtue of its distance from the Sun, perhaps it should be named for the Roman god of the dead.[6]

Fisher and Jayawardhana are both looking at stars similar to a young Sun, but that are as much as five times the Sun's mass. "We believe that if these disks have survived until the age of the stars we are looking at," says Fisher, then they "probably are similar to the disks that formed around the 'solar' system in which the planets were found. These are probably proto-planetary disks."

De Buizer, on the other hand, is looking at stars that are 20 to 30 times the mass of our Sun. There are far more low-mass stars than high-mass stars in the solar neighborhood, because low-mass stars live longer than high-mass stars (which tend to live fast and die young) and so have accumulated over the 10-billion-year life of our galaxy. Unlike Fisher, who picks his low-mass stellar targets from the old IRAS surveys, De Buizer uses methanol maser emissions from such high-mass stars to hone his list. De Buizer hopes to use masers to spot the tell-tale sign of dust disks around high-mass stars. "We know that masers trace denser material," says De Buizer. "This is one way of looking at these high-mass stars that have disks around them. It just so happens that methanol masers are excited around these very high-mass stars. We have not determined whether there are disks around these high-mass stars, but it's highly suggestive, since we see these linear-shaped, linear-distributed methanol masers, and we see elongations of dust in the same directions."[7]

MASERS
A maser (acronym for Microwave Amplification by Stimulated Emission of Radiation) occurs naturally in space any time radiation at a given frequency excites gas molecules sufficiently to emit further radiation along the same direction and wavelength. In 1965, radio astronomers detected the first known natural maser emission, at 1,665 megahertz. A synthesized maser was first created here on Earth in 1954, for the generation and amplification of microwaves.

By now it was midnight, the clouds were finally clearing over CTIO, and it was time to open the telescope's dome. Other astronomers were evicting us from our chairs to get at the computers and back to their work, as talk is cheap and a night at this telescope costs some $10,000. So, Fisher, De Buizer, and I took an elevator to the fourth level and walked up a metal stairwell to the base of the telescope. There, locked in for the night, was OSCIR, in a small metal cage below the telescope's primary mirror. As Fisher was giving me the rundown on how the telescope follows Earth's rotation and tracks its targets across the sky, the dome-slit above us whirred to life, startling us all.

Yet the two researchers seemed content, as with each passing month, new data supports the notion that while planetary disks may not be commonplace, in a galaxy as large as ours they are common enough to make the study of planet formation a life's work. A few moments after leaving them, walking outside into the pitch black, I felt my way along a gravel path and stopped to look up. There, at last, I could see the full galactic disk of the Milky Way, stretching from horizon to horizon, so thick with stars that I could only imagine the comparisons Huck and Jim might have made.

1 Fisher, Scott, graduate student, University of Florida. Interviewed on July 12, 1999, at the CTIO 4-meter telescope, Chile.

2 Blum, Jürgen, astrophysicist, University of Jena, Germany. Interviewed on November 25, 2000.

3 Gorkavyi, Nick, astrophysicist, NASA Goddard Space Flight Center, Greenbelt, Maryland. Interviewed on August 10, 2000.

4 McBreen, Brian, astrophysicist, University College, Dublin, Ireland. Interviewed on December 8, 1999.

5 Meyer, Michael R., astronomer, University of Arizona. Interviewed on November 28, 2000.

6 North, John 1994. *The Fontana History of Astronomy & Cosmology*. London: Fontana Press: 430.

   Leverington, David 1995. *A History of Astronomy: From 1890 to the Present*. London: Springer-Verlag: 97–98.

7 De Buizer, James, graduate student in astronomy, University of Florida. Interviewed on July 12, 1999, at the CTIO 4-meter telescope, Chile.

Spectroscopic Nights

CHAPTER 3

On a spectacular spring day in early May 1999, the domes at the Observatoire de Haute-Provence (OHP) stand gleaming through the scrub brush and small pines. All nine are scattered along the top of a small Provencal mountain, which, in truth, is little more than a long rolling hill within striking distance of the Luberon region in southern France. The air is redolent with lavender and thyme; evenings are almost perfect; the days clear and bright. But despite the languid atmosphere of the countryside, nothing can dissuade the French from following their rigid mealtimes. Their dinner bells could be used to calibrate sidereal clocks. So, like clockwork, Swiss astronomer Michel Mayor, Director of Geneva Observatory, and his wife Françoise arrive at the OHP's main dining room for lunch. There they take their places alongside a coterie of other astronomers, who are literally checking their watches.

I had first met Mayor in early 1996, two and a half months after he and his colleague Didier Queloz, a doctoral candidate at the time, had announced the indirect detection of a Jupiter-like planet circling 51 Pegasi—only three years after Wolszczan announced his discovery of the pulsar planets. Unlike Wolszczan, Mayor and Queloz had announced the discovery of a peculiar gas giant circling not a pulsar but a very normal Sun-like star almost 14 parsecs away from Earth in the northern constellation of Pegasus. Mayor and Queloz will always be remembered as the astronomers who discovered the first extra-solar planet in orbit around a Sun-like star. Yet for his part, other than a few more gray hairs in his beard, Mayor remains the same congenial, soft-spoken, and infinitely patient astronomer that I remember from our early 1996 meeting at his office outside Geneva. He is still sought out by the media, and this day was no exception, as he was being shadowed by a Montreal TV crew filming a documentary on planet hunting.

I knew very little about methods of extra-solar planet detection at our first meeting, so Mayor spent most of our time together, in between answering calls from other journalists, trying to explain Doppler spectroscopy to me. (This is a familiar role for Mayor, since he is also a professor of astronomy at the University of Geneva.) My only previous experience with spectroscopy involved an ill-fated effort to observe spectra emitted by various chemical elements and compounds in my high school chemistry lab. But from crude experiments like those routinely performed in almost any high school-level science lab, spectroscopy has grown into a tool that has, literally, opened our eyes onto the Universe.

BREAKING LIGHT INTO SPECTRA

Spectroscopy is crucial to astronomy. Without it, Mayor would never have been able to find the planet circling 51 Pegasi, now commonly referred to as 51 Peg. Almost everything we know about astronomy is somehow related to spectroscopy, including a star's composition, temperature, absolute magnitude (a star's true brightness), and luminosity. Spectroscopy is used to find new planets around other stars, to analyze comets and our own Solar System's planets, and to track galaxies and even the expansion of the Universe. In its most basic form, a spectroscope

Michel Mayor of Geneva Observatory on a platform outside the Observatoire de Haute-Provence's 1.52-meter telescope. (Photo by Bruce Dorminey.)

is nothing more than an optical device that breaks visual light into its component wavelengths, or parts, that appear as dense groups of spectra (lines) if projected onto a neutral background.

Spectroscopy owes its development to the work of German physicist and optician Joseph von Fraunhofer. By 1812, von Fraunhofer was busy designing telescope lenses, and was looking for a way to accurately measure how light refracts while moving through a telescope lens. By finding a standard and reliable wavelength measure against which to compare incoming light, he would be able to design a more up to date version of an achromatic lens (a lens that corrects incoming wavefronts of light for chromatic aberrations, or, essentially, false colors.) In 1814, von Fraunhofer used glass prisms to break down light from an ordinary yellow flame into spectra, or wavelengths. In the process, he created the first spectroscopes.

At first, spectroscopy relied on directing light through a slit-glass prism and a collimating lens, which aligned the light's rays so that they were parallel to each other before being focused at the eyepiece. Today, instead of prisms, spectroscopes use diffraction gratings. In astronomy, diffraction gratings are really "reflection" gratings—precision-etched, high-quality glass gratings that serve to disperse reflected light from celestial observations into its component wavelengths. Modern diffraction gratings are typically laser-cut, with at least 6,000 grooves per square centimeter.

Spectroscopy as a Tool for Astrophysics

Although von Fraunhofer used what would be considered very primitive glass prisms by contemporary standards, he was able to detect that sunlight emerging from his prism showed hundreds of dark lines. He mounted a prism on a telescope and measured the position of some 500 solar spectral lines, which he hoped could be used as wavelength standards. Unsure what they were, von Fraunhofer labeled them simply A to K, in decreasing wavelengths, beginning with the reddest. Then he began comparing light from his gas flame to light from the Sun, and even from other stars. He noticed similarities between the light from various salts burned in his flame and the spectra from sunlight.

In the late 1850s, some 30 years after von Fraunhofer's death, German chemist Robert Bunsen, working in concert with German physicist Gustav Kirchhoff, began experimenting with von Fraunhofer's methods in an effort to analyze salts. This marked the beginning of spectral analysis. They used an open-flame burner, later known as a Bunsen burner, to heat and analyze various salts for their signature emissions of color. By 1859, the two scientists had employed this method to identify the new elements of cesium and rubidium. The team soon began associating the colors with spectra obtained in sunlight, eventually enabling the identification of chemicals via their absorption lines. (Absorption lines are dark lines in a spectrum created by the absorption of radiation as it passes through a gaseous medium. In contrast, emission lines are bright lines given off by electrons of an excited gas, usually after being bombarded by radiation.)

In their visual form, spectral lines made from absorption show up dark against a bright background; in contrast, spectral lines made in emission are bright against a dark background.

Because the Sun's photosphere (surface) is much hotter and denser than its chromosphere (outer layer), radiation from the photosphere tends to be absorbed by the Sun's cooler outer layer. Proof of this comes through observation of a "forest" of dark spectral lines (sometimes referred to as Fraunhofer lines), which are revealed when placed against a neutral background. These spectral lines represent atoms and molecules that have been "excited" through collisions with other molecules, and that usually manifest themselves in a gaseous form. Such excitation normally is caused by thermal, electrostatic, or electromagnetic energy. Each spectral line represents a chemical compound, thus giving scientists the ability to identify the amount of and elements within any given spectrum.

Today, we know that von Fraunhofer's original D line represents absorption lines of sodium, while his H and K lines (two of the most prominent) denote absorption lines of calcium. By 1860, spectroscopes were well known among astronomers. Amateur English astronomer William Huggins mounted a spectroscope on his 8-inch telescope and used photographic plates to record spectra, thereby creating the first spectrograph. By 1890, 34 chemical species had been identified from solar absorption spectra, including gaseous forms of hydrogen, helium, calcium, iron, magnesium, carbon, and titanium. The science of astronomy moved into a new era.

The Measure of a Star

By 1918, some 225,000 stars had been measured and cataloged spectroscopically, in large part due to the generosity of Anna Palmer Draper, the widow of amateur astronomer and spectroscopist Henry Draper. Born in 1837, Draper was the son of well-known New York City physician John Draper. He followed his father's footsteps and by 1858 had graduated from the City University of New York with a degree in medicine. Draper spent time working at Bellevue Hospital and later became a City University professor of natural science, analytical chemistry, and physiology, but his first love was astronomy. His marriage to the daughter of New York City hardware and real estate tycoon Courtlandt Palmer enabled him to build a private observatory at Hastings-on-Hudson, New York, which saw completion in 1860.[1]

In 1872, on his observatory's 28-inch (71-centimeter) reflector telescope, Draper took the first photographic spectra of the bright summer star Vega—the same star that we now believe harbors a protoplanetary disk. Although Draper retired early in hopes of devoting his full time to astronomy, he died at only 45, a few months after his retirement in 1882, and before he was able to seriously begin cataloging spectra. His widow, Anna Palmer Draper, hoped that she could contribute to the continuation of his work in a meaningful way. At the behest of Edward Pickering, director of the Harvard College Observatory at that time, she established the Henry Draper Memorial Fund in 1886. With the assistance of Williamina Fleming, a young Scottish astronomer working at Harvard, Pickering oversaw work on the compilation and cataloging of several thousand different stellar spectra at the Harvard College Observatory.

By 1890, Fleming, Pickering, and their team of human "computers" (primarily women, who did the painstaking work of classifying reams of spectral data) had produced a first catalog of some 10,000 stars. Their new classification scheme, which was designed by Pickering and became known as the Harvard system, represented each stellar spectral type by a letter of the alphabet. At first, Pickering envisaged that the classification system would simply identify hottest to coolest, with A representing the hottest, and each subsequent letter representing a cooler type star. But this proved fallacious when it became known that O stars were hotter than A-type stars. By 1897, the order had been rearranged.[2]

Clearly, the true order of the stellar spectral range was uncertain, so understandably, before continuing the effort, Pickering wanted assurances that astronomers would accept Harvard's new classification system. Finally, in 1910, astronomers at the Bonn meeting of the International Solar Union voiced their approval, and the next year Harvard astronomer Annie Jump Cannon went to work on classifying 100,000 stars—again, with the aid of a group of human computers. They finished the initial list of 100,000 in two years' time. Buoyed by this success, they continued classifying additional stars for another two years, until 1915.

Between 1918 and 1924, the spectral types of 225,300 stars from all over the celestial sphere were published in nine volumes, titled the Henry Draper Catalog. (Extensions to the catalog have upped the original number to some 400,000, and it is now available via the Internet.) The basics of Draper's classification method are still in use (O, B, A, F, G, K, M designate the spectral types of main-sequence, or hydrogen-burning stars), but exist as a more refined system, called the Morgan-Keenan Classification System, first developed in 1943 at the University of Chicago's Yerkes Observatory.[3]

Thanks to Draper's initial classifications, we now know that the stellar spectral types O, B, and A are associated with the hottest and most massive stars in the galaxy. Due to their size and temperature, they end the stable hydrogen-burning phase of their lives quickly, often as supernovae. Consequently, O, B, and A stars are not considered ideal candidates for planet

A group of women computers at the Harvard College Observatory (circa 1890), directed by Mrs. Williamina Fleming (standing). (Harvard College Observatory.)

formation or for supporting planets that might harbor life. Their fast rotation also blurs their spectral lines, making high-precision spectroscopy difficult. And because they are so hot, many of the elements that show up in absorption spectra of cooler stars can't be seen in O, B, and A stars. F, G, and K stars, on the other hand, are more like our own. They rotate slowly and live longer, more stable lives.

M stars rotate even more slowly. They are the small red dwarfs most numerous in our "local neighborhood," out to 5 parsecs (16.3 light years). But they are also the most difficult to spot, because we do not live near a star cluster, at least not now. Our Sun probably formed in one, but that has yet to be proven definitively. Some star clusters are so dense that, if our Sun were part of one, there might be as many as a thousand stars between here and Proxima Centauri, our nearest neighbor at 1.3 parsecs. By contrast, our own neighborhood has a relatively low star density of only 0.14 per cubic parsec, 70 percent of which are small M dwarfs. G stars relatively similar to our own barely make a dent, at less than 10 percent of the local population.[4]

Stellar spectral types are further divided into numerical subspectral types, which range from 0 to 9. These numbers follow a descending mass-temperature curve: 0 stars are the hottest and most massive, while 9s have the least mass and are the coolest in any given subspectral type. Our Sun, a yellow G2 star, has a surface temperature of 5,770 degrees Kelvin. 51

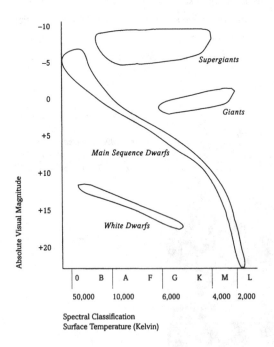

Spectral Classification
Surface Temperature (Kelvin)

This Hertzsprung-Russell diagram shows stellar spectral types arranged in a graph of luminosity plotted against temperature, classically done by using absolute visual magnitude (brightness increasing upward) and spectral class (temperature increasing leftward). The stars at the top are a million times the luminosity of the Sun, those at the bottom, a millionth. Stars separate into ordinary hydrogen-fusing main-sequence "dwarfs" like the Sun, huge giants, yet-larger supergiants, and tiny white dwarfs each kind in a different stage of life and death. Separate versions of this same diagram were developed independently in the early part of the last century by Danish astronomer Ejnar Hertzsprung and American astronomer Henry Norris Russell. Their ideas about how to chart stellar spectral types were subsequently combined into one diagram which shares their names. (Adapted with the Author's permission from The Little Book of Stars *by James B. Kaler, © 2001 James B. Kaler.)*

Pegasi, a G2.3 star, has almost the same temperature as our Sun. The smaller the mass and the lower the temperature, the longer stars remain stable. G stars, at the lower end of the spectrum, can burn for as long as 16 billion years. Radioactive dating of primordial materials from the early Solar System and lunar rocks tells us that the solar nebula spun down into a protostellar disk about 4.6 billion years ago, when it "switched on" and entered into its thermonuclear burning phase, fusing hydrogen into helium. By knowing the Sun's distance from Earth, and its gravitational effects on the planets in our own Solar System from their orbital periods, astronomers are able to determine the Sun's mass. By knowing its mass and calculating its surface temperature, which indicates its total energy output, physicists can then calculate the rate of reactions the Sun would need to achieve its current energy output. By applying a mathematical formula based on the Sun's current rate of hydrogen fusion reactions and its mass, physicists can extrapolate how quickly the Sun's hydrogen supply will be depleted. Numerical models give the Sun roughly another 6 billion years in its current stable hydrogen-burning state, which means that our Sun is almost halfway through its estimated 11-billion-year life as a main-sequence star. This main-sequence phase will end when the Sun goes into its next phase as a red giant, which begins when the Sun has used up its core supply of hydrogen and is forced to start burning helium.

SOLAR FUSION AND STELLAR EVOLUTION

In 1938, the German physicists Hans Bethe and Carl Friedrich von Weizsäcker, working independently, theorized that stars release nuclear energy by converting hydrogen into helium, in what became known as the carbon-nitrogen cycle (or the Bethe–von Weizsäcker Cycle). This cycle takes place in the form of thermonuclear fusion, with carbon, nitrogen, and oxygen acting as catalysts in speeding this six-stage reaction. Proof of the power of such fusion came 14 years after Bethe and von Weizsäcker proposed it, in the form of an atomic test, whose own catalyst was the Cold War. That's when, in 1952, the U.S. exploded the world's first hydrogen bomb on Enewetak Atoll, a chain of islands near the Marshall Islands in the central Pacific. This new H-bomb generated its energy from fusing two isotopes of hydrogen into the heavier element of helium. A fission trigger was used to generate the extremely high temperatures needed to fuse deuterium with tritium, and tritium with tritium, in a locally devastating demonstration of the Sun's source of energy. (The bomb completely obliterated one of the atoll's smaller islands.) The fledgling computer power necessary to run the H-bomb calculations led to a whole new industry, which blossomed in the form of supercomputers in the 1960s. Such supercomputers, in turn, enabled astrophysicists to run numerical stellar models on the lifetime of the Sun. Masses of other stars, an estimated 60 to 80 percent of which are binary (or double) stars, can be determined using Doppler spectroscopy, whereby the same principle used to determine the rough mass estimates of planetary companions is applied to determine the mass of any given binary star's stellar companion. Thus, by determining a star's mass, coupled with its spectral type and color, which reveals its temperature, astronomers are able to determine estimated lifetimes for the full range of stars in any given star catalog.

STELLAR VELOCITIES

Lying in a meadow on a lazy summer afternoon, few of us would interrupt our daydreams to consider that we were actually hurtling through the void of spacetime at 30 kilometers per second. But motion and velocity (the speed at which a body moves in any given direction) are essential to extra-solar planet detection.

Determining velocity is inherently subjective and is very much dependent on the eye of the beholder. Thus, we can only estimate Earth's velocity through spacetime relative to the movement of the galaxy as a whole, and relative to our own local stellar neighborhood. However, some planet hunters use velocity measurements to register a perceptible "wobble" caused by the gravitational pull an extra-solar planet exerts on its parent star. For example, as viewed from a distance of 10 parsecs from our Solar System, the gravitational orbital influence of Jupiter would be observed to cause the Sun to be jerked around its center of gravity by 500 microarcseconds (100,000 kilometers). By contrast, Earth's own gravitational perturbation on the Sun (again as seen from 10 parsecs away) would only cause the Sun to jitter by 0.3 microarcseconds, a scale equal to that of resolving the diameter of a human hair at a distance of 30,000 kilometers. Such motion is the key to Doppler spectroscopy, which to date has been the most successful method of extra-solar planet detection.

MICROARCSECOND
A microarcsecond denotes a measure of an angle, which in an astronomer's case would be the measure of angle between two bodies on the celestial sphere. The whole celestial sphere is divided into 360 degrees of arc, with each degree equaling 60 minutes, and each minute equaling 60 seconds of arc. One arcsecond is then further divided into hundredths, thousandths, and millionths of a second, with one microarcsecond equal to one-millionth of a second of arc.

Dawn of Doppler Spectroscopy

Austrian physicist Christian Johann Doppler spent the first part of the nineteenth century trying to understand why some stars are of different colors than others. He hoped to observe changes in star color caused by alterations in velocity that would correlate to observable red and blue shifts of visible light. In the process, he discovered the physical principle now known as the Doppler effect, which laid the groundwork for Doppler spectroscopy, giving astronomers the ability to measure radial velocities, or velocities along our line of sight.

Although Doppler's ideas about star coloration were eventually cast aside, the terms redshift and blue-shift stuck. Red-shift refers to the fact that red light corresponds to longer wavelengths (at the lower end of the visible spectrum), whereas blue-shift indicates that blue light corresponds to shorter wavelengths (toward the upper end of the visible spectrum).

STAR COLORATION
Stellar spectra exhibit what are known as red-shifts and blue-shifts. These effects are best observed with sophisticated spectroscopy. Star coloration is caused either by atmospheric interference here on

Earth or by astrophysical processes within the stars themselves; it is not due to their relative velocities or motions through the heavens.

At higher frequencies in the electromagnetic spectrum (from infrared light to visible light to gamma rays), astronomers use Doppler's principle to record radial velocities of planets, stars, and galaxies as these celestial bodies shift toward or away from us, along our line of sight. As light or other emitted electromagnetic radiation from a body moving toward us is observed here on Earth, it appears to be shifted toward the higher electromagnetic frequencies. This shift is observed as compressed (or blue-shifted) electromagnetic radiation relative to our own point of observation. If the object emits radiation while moving away from us, then its spectra appear to be elongated or red-shifted away from us. The degree of such shifts can be used to estimate the speed of a body's motion toward or away from us along our line of sight.

Fiducial Requirements

To obtain Doppler spectroscopy measurements, astronomers first have to record an object's radial velocity, or its rate of movement along our line of sight. Stellar spectra from slowly rotating stars can provide dense forests of thousands of hydrogen spectral lines that can be used to precisely measure a wavelength's motion toward or away from us. To make the technique work, astronomers must have what is known as a fiducial (or fixed base of reference) in order to compare the stellar spectra against spectra from a stable source. The key for any fiducial, though, is that it be made stable enough so that it may be trusted as a zero-velocity control against which to measure stellar spectra with high precision.

If astronomers record a positive velocity in the wavelengths of the spectra, it means that the star is moving away from us, or red-shifting. If it is negative, it means that the star is blue-shifting, moving toward us. Astronomers have been using crude variations of this Doppler technique to measure the velocities of distant galaxies since 1912, even before they were known to be galaxies. Over the decades, there have been numerous refinements of such measurement systems. For example, measurements showing galaxies moving away from us at high red-shifts provided cosmologists with the first widely accepted clues as to how fast the Universe is expanding. But it wasn't until the late 1970s that spectrographs became precise enough to enable measurements of perturbations caused by small substellar objects, be they binary companions or a Jupiter-mass planet. During that decade, Gordon Walker, Bruce Campbell, and Stephenson Yang, all formerly of the University of British Columbia in Vancouver, began experimenting with Doppler spectroscopy at Canada's Dominion Astrophysical Observatory in Victoria, British Columbia.

Because of the role Walker played in honing Doppler spectroscopy as a means of finding planets, I went to meet him at one of Victoria's harbor-view hotels less than two weeks after I had shared lunch with the astronomers at the Observatoire de Haute-Provence. Gesturing across a cocktail table, Walker explained how a fiducial is incorporated into planet detection: "If you and I were separated by a cloud of hydrogen, the bright light would have a series of dark lines across it, due to ion transitions [molecules in an excited state of flux caused by the loss or gain of electrons] in the hydrogen. Because of this, there are a lot of water vapor lines that arise naturally in Earth's atmosphere. In 1972, using Earth's atmosphere as a fiducial, John Glaspey, at Kitt Peak Observatory near Tucson, Arizona, looked at Algol [a variable star] with very high precision. Because Algol is part of a multiple star system, Glaspey was able to detect very small velocity changes in that star's hydrogen lines. And it just came to me: 'If we can do that, then we can start looking for planets via the radial velocity perturbation technique.'"[5]

COMPETING METHODS

Prior to the development of Doppler spectroscopy, the only other reliable way to look for extra-solar planets was to watch for tiny perturbations in a star's proper motion—that is, the way it moves across our line of sight, a process called astrometry (discussed more fully in Chapter 10). Planet detection via astrometry is, however, a precarious and painstaking technique that takes years, and its path is littered with countless false detections. By contrast, Doppler spectroscopy measurements can be used in some cases to detect extra-solar planets in less than a week. In Doppler spectroscopy, the only thing that matters is that a star changes its velocity from one observation to the next. If it does, and with an ascertainable periodicity, a likely conclusion is that the star is experiencing what's called a barycentric reflex motion (around its barycenter, or center of mass) due to the orbit of either a stellar mass or substellar mass companion.

Bruce Campbell, Walker's postdoctoral researcher, proposed taking Glaspey's idea a step further. He suggested that, instead of measuring distant stellar spectra against the hydrogen lines in Earth's atmosphere, it would be easier to use a captive gas in a container attached to the "skirt" of the spectrograph. The dark absorption lines of the zero-velocity gas could be directly imposed onto the incoming stellar spectra. The question was which gas to use for a fiducial. For the answer, the Canadian team went to Gerhard Herzberg, a Nobel Prize-winning

Gordon Walker, who is now "officially" retired from the University of British Columbia in Vancouver, is still active in observational collaborations, as indicated by this recent photo taken in the control room of the Canada-France-Hawaii Telescope atop Mauna Kea, Hawaii. (Photo by Dr. John Maier, University of Basel.)

chemist at the Canadian National Research Council's Herzberg Institute. Herzberg and his colleague, Alex Douglas, suggested using hydrogen fluoride, a very powerful acid that is lethal in large quantities. It proved to be a remarkable fiducial. The hydrogen fluoride spectrograph made it possible to take Doppler velocity measurements at an accuracy of only 13 meters per second, or about the speed limit of a car in a residential zone. (Even so, 13 meters per second was a far cry from Alexander Wolszczan's radiopulsar-timing measurements of 1 meter per second, a pace easily beaten by even the slowest riding-lawnmowers.) In theory, this new accuracy would afford the Canadians the ability to detect large planets orbiting nearby normal Sun-like stars.

Two days after Christmas 1978, using the Dominion Astrophysical Observatory's 1.2-meter telescope in Victoria, Walker and Campbell made the first observations using a hydrogen fluoride fiducial. From 1980 to 1992, the planet-hunting team of Walker, Campbell, and Yang took data from 21 stars, using both the telescope in Victoria and the 3.6-meter Canada-France-Hawaii Telescope (CFHT) atop Mauna Kea on the big island of Hawaii. The team began to achieve a yearly average of 30 to 40 observations per star. By 1988, they were reporting that Gamma Cephei, a K star some 16 parsecs from Earth, was showing velocity changes that mimicked the gravitational perturbations that would have been caused if the star were being circled every two and half years by a Jupiter-like planet. But, in fact, Gamma Cephei turned out to have significant variations in its interior stellar processes. Its apparent velocity fluctuations were not due to planets circling it in periodic orbits, but to periodicities in its own energy output. This, in turn, caused the star's outer chromosphere to expand and contract, thus giving the planet-hunters radial velocity readings that made it appear as if the star itself were being shifted to and fro by the gravitational tug of a nearby planet. Even our own Sun, though completely stable, has regular cycles of variability, as evidenced by a significant increase in the number of sunspots, or strongly magnetic cool areas near the solar surface, which can emit high-energy solar flares. Every 11 years, almost like clockwork, Earth suffers a maelstrom of solar activity that wreaks global telecommunications havoc. Even stars that aren't variable may have an active upper layer, or chromosphere. Chromospheric activity can also mimic the Doppler spectroscopy perturbations caused by a real planet. The Canadians' measurements of Gamma Cephei were indicative of exactly this kind of activity. Clearly, planet hunters using the Doppler spectroscopy method must always measure the stars' chromospheric activity to make sure that the periodicity that they detect in the perturbations of their spectral lines are caused by a planet and not by a star's surface activity. "In the Sun," says Walker, "it takes about a million years for heat generated in thermonuclear reactions to diffuse to the surface. It is rather like a bullwhip: if you flick it slightly, the wave gets larger and larger in amplitude, until it cracks and exceeds the velocity of sound."

While the Canadian team was mired in the vagaries of stellar physics, Michel Mayor, together with the late Antoine Duquennoy who was his colleague at the Geneva Observatory, was continuing what would ultimately be a 13-year systematic survey of all Sun-like (F7–G9) stars out to 22 parsecs, in an effort to determine the frequency of binary, or double, stars. (Our own Sun rests among the minority of stars in the local neighborhood that is not a part of a binary system.) For the survey, Mayor and Duquennoy chose to use a collection of old stars from the Gliese star catalog.

THE GLIESE STAR CATALOG

The Gliese "Catalog of Nearby Stars" gets its name from the late German astronomer and physicist Wilhelm Gliese (1915–1993). "Mr. Nearby Stars," as Gliese was sometimes affectionately called, began studying stars in our local neighborhood in the early 1950s. His now-famous catalog, which lists stars all roughly within 20 parsecs of Earth, was a natural progression of his interest in our stellar neighborhood and is still used nightly by stellar astronomers. Most of these stars are now referred to solely by their Gliese catalog number, e.g., Gliese 229.

Mayor, who had written his doctorate on the spiral structure of galaxies, came to realize that the kind of kinematic data he needed was very rarely found in existing catalogs. To meet his need, he and colleagues partnered with the Marseille Observatory in developing the Correlation and Velocity (CORAVEL) spectrograph, which was, in turn, installed at Observatoire de Haute-Provence. At an initial accuracy of 250 meters per seconds (considered excellent when the survey began in the late seventies), CORAVEL enabled Mayor to conduct surveys on the dynamics of globular clusters, groups of very old stars in our galaxy's halo. However, the binary survey was conducted not only to detect binary star systems, but also to establish parameters on their orbital dynamics. That led Mayor to think about how to detect even lower-mass stellar companions.

PLANET HUNTING AT OHP

Mayor and his Swiss and French colleagues soon realized that the lower they could hone their Doppler spectroscopy accuracies, the better chance they would have of detecting planets around other stars. And because Observatoire de Haute-Provence is only a four-hour drive from Geneva Observatory, and the French have many more clear nights than the Genevese, it was only logical that the Swiss would eventually build their own guest residence at OHP. So, only a couple of hours after having lunch with Mayor and his colleagues on that same spring afternoon in May 1999, he and I sat down for a leisurely chat in the small study of the guesthouse. "With CORAVEL," explains Mayor, "we had made many discoveries in the range of brown dwarfs [substellar-mass failed stars], and we knew that these kinds of objects were essentially rare; so we knew that if we wanted to make progress in that field we had to have a significant sizeable sample. Therefore, we decided to look at a larger sample."[6]

Mayor and Queloz selected 142 of the stars cataloged from Mayor and Duquennoy's binary survey, stars that were known to be nonbinary and nonvariable. They also decided to use a new higher-precision spectrograph, the product of a joint OHP-Swiss collaboration, dubbed ELODIE in honor of an OHP secretary. Using ELODIE would also enable them to perform online data analysis at the dome. "When we asked for time here in 1993," Mayor told me, "our search was aimed at substellar companions, not specifically brown dwarfs or extra-solar planets, but companions below the hydrogen-burning limit. Nobody knew what kind of orbits these brown dwarfs would have, so we were open to a low-mass object with a short period." And therein lies the secret to their success: while their competitors in the U.S. and Canada were looking for longer-period orbits, Mayor and Queloz were on the lookout for anything, even a planet on a shockingly short orbit.

In contrast to Walker, Mayor's team uses emission lines from a thorium calibration lamp mounted just below and parallel to the telescope's primary mirror. (Thorium is a radioactive metal used as fuel in certain nuclear reactors, but in this case is in gaseous form.) Two fiber-optic cables link the telescope to a thermally controlled room below the dome, where a spectrograph records light coming from the star and from the lamp itself, then mixes and compares the two in a live feed, which appears on a computer monitor in the control room. The measurements are computed while the stellar image is on—that is, "live."

The image on the computer screen displays both the starlight and the lamplight. Horizontal lines indicate the star's spectrum; the tiny dots in between are from the emission lines of the thorium calibration lamp. In Walker's hydrogen fluoride spectrograph, his fiducial was imposed directly on incoming starlight. The hydrogen fluoride thus created absorption lines as the starlight passed through its gas. (Walker has since retired, so the disagreement between the two scientists over which is the best method will never be resolved.)

Ultimately, ELODIE enabled the detection of 51 Peg's variable velocity. It revealed that the star is moving directly toward us at −33 kilometers per second. When 51 Peg was first detected, ELODIE's accuracy rate was down to +/− 12 meters per second. The planet itself causes a reflex motion (a gravitational tug or jerk) on its parent star by +/− 59 meters per second. To visualize this, imagine children on a small merry-go-round, of the sort usually found on a school playground, only two or three meters in diameter, and equipped with crude metal handles for grips. To someone sitting on a park bench focusing only on the merry-go-round's center stanchion, the stanchion will appear to jerk toward or away from the viewer along his or her line of sight. The larger the child riding along the merry-go-round's outer edge, the more the stanchion will move. By watching only this central movement, an observer can make some deductions about the child's mass and size. Astronomers follow the same basic principle when using Doppler spectroscopy to hunt planets.

After dinner that night, Mayor drove up a short hill to the 1.93-meter telescope, the one on which he and Queloz discovered the planet around 51 Pegasi. He was there to oversee the last of a seven-night observing run. Usually, aside from the distant drone of a passing car, OHP and most other observatories like it are deafeningly quiet. Most observational astronomers— at least those who don't discover planets—seem akin to scientific monks, as they spend much of their time cloistered on acres of secluded land, hunched over PCs. But on this night, the visiting Montreal television crew had turned the telescope's control room into an impromptu stu-

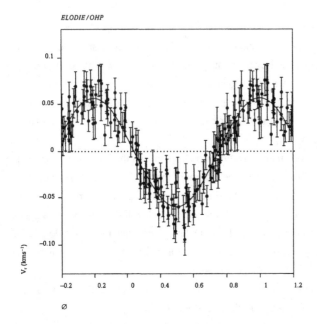

This data graph shows the periodicity in the radial velocity curve that is the telltale signature of a planet orbiting the star 51 Pegasi every 4.23 days. (Diagram by Didier Queloz.)

dio, which they commandeered until 10:00 P.M., when Mayor and I were finally free to continue our conversation in peace.

"51 Peg's first sign of velocity variability was evident in the fall of 1994," says Mayor. "By January 1995, we already had a good ephemeride [estimation of the suspected planet's movements] to predict the future velocity of the star." But before Mayor and Queloz could start celebrating, they were forced to take a six-month hiatus, as the star was changing its position in the sky due to Earth's orbit around the Sun. It wouldn't be visible again until the following July. It wasn't until July 3, 4, and 5, 1995, that they could verify their ephemerides for the star. "We checked that we were following what was predicted," said Mayor, "and we checked that the detected variability was still here with a good phase, amplitude, and good period. We had a very strong argument that it was not an intrinsic variability, but a stable phenomenon, like planets. Then we really started to be enthusiastic."

They concluded that 51 Pegasi, a G2.3 star 13.7 parsecs away from Earth, was being orbited by a planet about half the size of Jupiter every 4.23 days—a distance some eight times closer to its star than Mercury is to our Sun. Mayor and Queloz first announced their finding at a conference in Florence, Italy, on October 6, 1995. (Subsequently, they published a peer-reviewed, refereed paper on the find in the November 23, 1995, edition of *Nature*. Alan Boss, a planetary theorist at the Carnegie Institution in Washington, D.C., Adam Burrows, a planetary scientist at the University of Arizona in Tucson, and Gordon Walker were the referees. At the time of the paper's publication, Walker was the only one of the three referees who refused to approve the finding, although he would later acknowledge the validity of the Swiss discovery.)

"I was completely astonished and surprised in Florence," Mayor remembers. "We received faxes at the hotel from all over the world, asking for interviews." Queloz adds, "When we announced 51 Peg, no one was expecting this kind of object, and we got two reactions. Some said, 'Okay, that's how science goes; I don't care if the theory works, this is new stuff and we have to deal with it.' Others said, 'No, this cannot be true because we know how a planet has to be formed—you've found something else.' But the data tells me that it is a low-mass object, and I don't care how it's formed."[7] Detractors had questions about the actual mass of a planet in such close orbit around its star, for Doppler spectroscopy has an inherent major pitfall: it cannot give hard-and-fast data on the mass of the planets it has detected. The reason is that observers may not be seeing the gravitational effects of the planet edge-on as it orbits its parent star. The Doppler spectroscopy technique is most sensitive to planets tugging at a star as it travels in front of and behind the star along the telescope's line of sight. If a planet is inclined to its parent star, then the extra-solar planetary system will appear to be more face-on to our line of sight. As a result, it is not possible to see its full gravitational perturbation on the parent star. Mayor says that, on statistical grounds, a planet's true mass should on average not be more than 30 percent above the lower limits that are quoted as its mass.

"My initial reaction," says Walker, "was that they were not seeing a dynamical effect, but rather chromospheric activity. The effect of such a short-period planet stuck out like a sore thumb; it was a huge perturbation. You don't require the same level of precision that we had been going for, which had been based on a 10-year orbit and a 10-meter-per-second amplitude. Here they've got a huge amplitude over a few days. Plus, the initial paper was based on data

Didier Queloz of Geneva Observatory is the co-discoverer of the planet that circles 51 Pegasi, the first known planet to orbit another Sun-like star. (Photo courtesy Didier Queloz.)

taken over two years, not on a period of four consecutive nights. So I was concerned that [Mayor and Queloz] might have an artifact, an aliasing of a longer period into a short period. I simply said this was too important to have another false alarm. The history of planet searches is full of false alarms." Yet within a few weeks, Walker had changed his mind, and Mayor and Queloz had company. Two San Francisco-based astronomers had not only confirmed the Swiss discovery, but had announced similar planet detections of their own using almost the same technique.

1   Daintith, John et al. (eds.) 1994. *Biographical Encyclopedia of Scientists* 1, 2nd ed. Bristol & London: Institute of Physics Publishing: 237.

2   Leverington, David 1995. *A History of Astronomy: From 1890 to the Present*. London: Springer-Verlag: 127.

3   North, John 1994. *The Fontana History of Astronomy & Cosmology*. London: Fontana Press: 423–424.

     Leverington, David 1995. *A History of Astronomy: From 1890 to the Present*. London: Springer-Verlag: 4.

4   North, John 1994. *The Fontana History of Astronomy & Cosmology*. London: Fontana Press: 483.

5   Walker, Gordon A. H., astronomer, University of British Columbia. Interviewed on May 16, 1999, in Victoria, Canada. A follow-up took place on January 15, 2000.

6   Mayor, Michel, astronomer, Geneva Observatory. Interviewed on May 6, 1999, at Observatoire de Haute-Provence, France. A follow-up took place on September 8, 2000.

7   Queloz, Didier, astronomer, Geneva Observatory. Interviewed on May 21, 1999, at Flagstaff, Arizona. Follow-ups took place on November 30, 1999, and June 5, 2001.

Planetary Hype
CHAPTER 4

The same week that most everyone else seemed to be watching Will Smith battle locust-like aliens in the July 1996 film *Independence Day*, I was battling several Italian taxi drivers in an effort to get a reasonable fare from the train station to the Naples ferry dock. Astronomers have a knack for picking eye-popping conference locales, so, despite transport hassles, I was certainly glad I had taken Michel Mayor's advice to attend the Fifth International Conference on Bioastronomy on the Italian isle of Capri.

During the day, hordes of tourists arrived to window-shop in the old town's main square. But after 10:00 P.M., the plaza's relative quiet almost made it possible to follow the shifting opinions of several cappuccino-sipping planetary theorists. It was here in Capri that I first met the planet-finding team of Geoffrey Marcy and Paul Butler, both then of San Francisco State University. I had happened upon Butler early one evening on the terrace of my small hotel, where he was enjoying a plate of pasta in relative anonymity. By the spring of 1996, Marcy and Butler's work had already been touted on the cover of *Time*, they had been interviewed on

ABC's *Nightline*, and they had been made ABC News "Persons of the Week." Now, complained Butler, they were being stalked by a crew that was filming a Discovery channel documentary.

Wolszczan and his discovery of pulsar planets had barely registered on the media's radar screen. Likewise, the Mayor-Queloz announcement of the planet circling 51 Pegasi had largely been ignored by the American media. But Marcy and Butler had, independently, confirmed a planet around 51 Pegasi, and had begun to make new planet detections of their own around the stars 47 Ursae Majoris and 70 Virginis. So they were now supplying the heat behind the media's latent "planet fever."

But Mayor and Queloz's find had taken Marcy and Butler by surprise, sending the California duo scurrying back to massive sets of heretofore unanalyzed data in search of planets in short orbits. They looked through reams of material on stars that they had been surveying since 1987 at the University of California's Lick Observatory. There, they opted not to use the Swiss thorium technique, but rather to follow the Canadian lead—to impose the fiducial on the stellar spectra in absorption. However, Marcy and Butler chose iodine gas over hydrogen fluoride. Iodine, the Americans found, was less reactive than hydrogen fluoride, and consequently less dangerous.

Marcy and Butler were still kicking themselves, wondering how they could have missed 51 Pegasi. They had left it off their observing program because they had mistakenly believed it to be an evolved, old star, meaning that it would be inherently unstable and therefore not lend itself to planet detection via Doppler spectroscopy. Mayor, though, had correctly surmised that 51 Pegasi might very well still be a stable main-sequence star, hence be a good candidate for planet detection via spectroscopy.

"I don't think of Michel Mayor as a competitor," says Marcy, with several years of hindsight. "I think of him as a wonderful international colleague. He's doing the best work aside from us, and I don't ever doubt anything he says."[1] That may all be true. But it's the sort of mag-

Prolific planet finders Geoffrey Marcy and Paul Butler pictured here in the control room of the Keck I telescope atop Mauna Kea in Hawaii. (Photo courtesy SUN Microsystems.)

nanimous post-game locker room talk that flows easily when your team has discovered more planets than anyone anywhere. At last count, Marcy and Butler's team had detected more than forty.

PLANETARY BONANZA

Still, as Marcy is quick to point out, it was the Swiss team that first opened the world's eyes to how different and special our own planetary system is from the fledgling Jupiter-like bodies that have been found circling in such close orbits. With that in mind, Marcy and Butler proceeded to blow away the astronomical community with discoveries that showed that Mayor and Queloz had only cracked open the door on an entirely new—and what could only be described as wacky—breed of planets. By 1996, Marcy and Butler had begun to find supermassive gas giants in very bizarre orbits around what were essentially very ordinary Sun-like stars. They announced planets around 70 Virginis, a G5 star, some 18 parsecs away in the Virgo constellation, and 47 Ursae Majoris, a G0 star, slightly warmer than our own Sun, some 14 parsecs away in the constellation of Ursa Major. The latter is one of the stars that make up the Big Dipper: 47 Ursae Majoris b, as its 2.4-Jupiter-mass companion is called, has a three-year orbit. Among the planets discovered at such relatively long distances from their parent star, this one has an uncharacteristic, nearly circular orbit. 70 Virginis b, by contrast, has a mass of around 8 Jupiters and orbits its parent star every 117 days in an eccentric elliptical orbit.

EXTRA-SOLAR PLANETARY NOTATIONS
The International Astronomical Union (IAU), a Paris-based organization which, among other things, oversees nomenclature for astronomical bodies, has yet to approve a final scheme for the formal notation of extra-solar planets. To date, there are several different methods of officially notating these substellar mass objects, but for the most part, planet hunters have taken their cue from astronomers who specialize in the study of binary and multiple star systems. With what are known as visual binary stars, where both components can be detected via a telescope in optical wavelengths, usually the brightest of the two stars is designated with an upper-case A following its proper name or catalog name, e. g., 16 Cygni A and 16 Cygni B. However, unseen stellar companions that have been detected via indirect means, such as Doppler spectroscopy or astrometry, are designated with a lower-case letter in the order of their discovery. Thus, the planet around 51 Pegasi is designated as 51 Pegasi b, while the planet around the B component of the binary star system 16 Cygni A and 16 Cygni B is designated 16 Cygni Bb.

Extra-solar planets are generally designated hierarchically, in the order of their discovery, with their designation beginning with the second letter of the alphabet, in deference to the parent star which is understood to be the A (or primary) component of the system. In binary star systems, as noted above, even if the star itself is designated as 16 Cygni B, its first discovered unseen companion would still be designated as b. Because the outer companion to Gliese 876 was discovered before its inner companion, it is known as Gliese 876 b, while the inner companion is designated as Gliese 876 c. With competing systems still in use, this whole issue of nomenclature for stellar and substellar companions is even confusing for astronomers. The matter may be resolved in 2003, when the IAU has its next General Assembly.[2]

Before the week was out in Capri, Marcy and Butler had made their first public announcement of a Jupiter-mass planet in a short 4.6-day orbit around Upsilon Andromedae. This F8 star some 13.47 parsecs away from Earth in the Andromeda constellation is clearly visible to the naked eye during summer and autumn. "Hot Jupiters," as they were now being called, were becoming a phenomenon themselves. But little did anyone know that Upsilon Andromedae, included in a Lick Observatory survey of 107 stars, also harbored two other planets, which would make it the first known multiplanet system circling another Sun-like star.

Marcy and Butler seemed to be founding members of a planet-of-the-month club. As the sheer numbers of new planets added up, so did the number of their partners in the search: Robert Noyes at the Harvard-Smithsonian Astrophysical Observatory; Steve Vogt, a longtime planetary theorist and astronomer at the University of California at Santa Cruz; Timothy Brown at the High Altitude Observatory in Boulder, Colorado; and Debra Fischer, Marcy's postdoctoral researcher at the University of California at Berkeley. (At the time of this writing, Marcy and Butler continue to cooperate on their planet surveys, though Marcy is now a professor of astronomy at Berkeley and was picked by the California Science Center as the 2001 "Scientist of the Year." Butler has taken a full-time research position in the renowned Department of Terrestrial Magnetism at the Carnegie Institution in Washington, D.C.)

Quieting the Noise

In the months following the announcement of Upsilon Andromedae b, it became clear that there was a lot of "noise" in the data, noise that couldn't be resolved by chromospheric emissions (or variability from the star). So Marcy and Butler spent three more years conducting extensive observations, which led them to conclude that the star's sole inner planet had at least one, maybe two, companions.

Armed with years of independent data from both the 0.6- and 3-meter telescopes at Lick Observatory near San Jose, California, and the Smithsonian's 1.5-meter telescope at the Whipple Observatory near Tucson, Arizona, they called a press conference in San Francisco in mid-April of 1999 to announce their findings—that indeed Upsilon Andromedae harbored a full-fledged planetary system. About halfway through its 6-billion-year life as a main-sequence star, Upsilon Andromedae, they reported, was being circled by two additional companions, meaning there were at least three known planets in the system. The middle planet, about twice the size of Jupiter, is in a 241-day orbit, and an outer planet of at least 4.3 Jupiter-masses is in an approximate 3.5-year orbit. The Upsilon Andromedae system is very different from our own, most notably because its three large planets are crowded into eccentric orbits comparatively close to their parent star. Imagine it this way: if the Upsilon Andromedae system were placed on top of our Solar System, its middle planet would lie somewhere between Venus and Earth, and its outer planet would lie in the asteroid belt between Mars and Jupiter. But none of them comes close to resembling the actual Jupiter, which unfailingly circles our own Sun every 11.862 years.

If the close-in so-called "hot Jupiters" were bizarre, then this weird system of three Jupiter-mass planets where the two outer bodies circle their star in eccentric, oval-like orbits, really presented a conundrum. And the team warned that because of the system's eccentricities, the planets might orbit the star in a non-coplanar fashion, more like electrons circling the nucleus of an atom or bees buzzing around a hive, than planets on orderly co-planar orbits around a parent star. "Paul and I have had enormous luck," says Marcy, who despite the accolades has kept a level head. "We're not out for glory. We're just out to answer a question I asked myself when I was 14 years old and looked up at a poster of our own Solar System: namely, are there other planetary systems out there? I feel lucky I'm even sitting here talking about it. I could have easily floundered in school. Even to this day, I wonder if I'm really smart enough to be contributing at all." At the 1999 Bioastronomy conference, which met in August on the big island of Hawaii, Marcy presented the latest data on Upsilon Andromedae. (Butler took the role as lead author when the team published its research on the system in *The Astrophysical Journal*, December 1, 1999.)

While in Hawaii, Marcy also took the time to give me an impromptu lesson on how to use a diagram of velocity versus time in order to see the different amplitudes of the three separate planets as they orbit their star. "With Upsilon Andromedae," said Marcy, pointing to his graphs, "you can see a clear periodicity of 4.6 days, but the scatter [or variety of the data's positions as plotted on the chart] is enormous. We were worried that we had somehow made a mistake. This is the original data set, velocity versus time. You can see the points going back to 1992, and your eyeball sees no periodicity there. So I subtracted the 4.6-day periodicity. Now you see the data points minus that sinusoid [wavelength curve] of velocity versus time. We tried a model with one additional planet: it absolutely doesn't work. So we were forced to try a model with two additional planets, and here it is: there is the long-term period of 1,300 days, and the

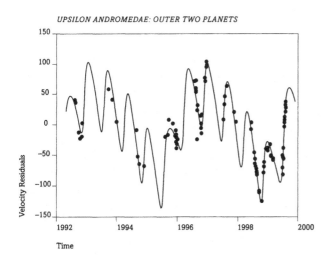

UPSILON ANDROMEDAE: OUTER TWO PLANETS

Radial velocity curve indicating the two outer planets orbiting the star Upsilon Andromedae. With data from the previously reported inner planet removed, periodicities indicative of two outer planets easily fit the expected radial velocity curves, which are plotted here over an eight-year time frame. (Geoffrey Marcy, U. C. Berkeley.)

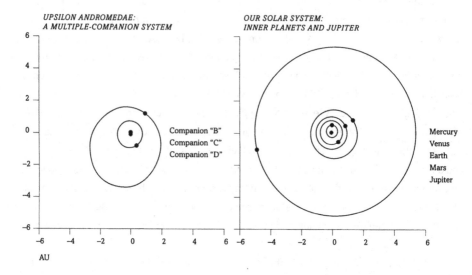

These figures show the relative differences between the orbits of the three known planets around Upsilon Andromedae and the orbits of planets in our own Solar System. (San Francisco State University.)

intermediate period. Having subtracted this short-period 4.6-day sinusoid, we can construct a model that has two additional planets. And that model with two additional planets fits exquisitely. It's quite remarkable. We have no choice but to conclude that a model with three orbiting objects best explains the velocity variations in Upsilon Andromedae."

THE A-LIST

Some three months later, by the end of November 1999, there were hints that two other stars, HD 217107 and HD 187123—each of which had been reported a year earlier as harboring single planets—might also have additional planetary companions. Kevin Apps, at the time a third-year astrophysics student at the University of Sussex in Brighton, U.K., was at least in part responsible for the initial discovery of both planets. In a cheeky move, he had E-mailed Marcy asking to see his target list. Apps was convinced that Marcy and Butler were targeting stars in searches that would never yield results, a claim based on data he had garnered from the European Space Agency's Hipparcos star catalog. (The catalog had been produced from observations made by the Hipparcos satellite over a four-year period ending in 1993, as part of the first comprehensive space-based effort to survey stars to determine their exact positions, proper motions, and distances from our Sun.)

"I realized that about 30 [of the stars] were unsuitable," says Apps, "because [Marcy and Butler] had based their picks on old catalogs, not the Hipparcos data."[3] Though Apps never really expected it, Marcy did in fact reply to his E-mail request and supplied Apps with their complete list of targets. With confidence belying his age and experience, Apps then wrote back, suggesting that Marcy and Butler delete 30 from their list and replace them with 30 of his own. As it turned out, Apps' original list of 30 alternative stars would prove to be directly responsible for the discovery of two new possible planetary systems. Subsequently, Marcy's team (by now being referred to as the California Group) put Apps in charge of picking targets. "I'm responsible for determining the target star's chemical abundances and spectral types," says Apps, who is now listed as a collaborator in the detection of an additional three planets.

The year 1999 would end on an even more tantalizing note, with the possible observation of the first direct reflected spectra from one of the hot Jupiters—also referred to as "roasters" because of their short orbits. For example, in 1997, Marcy and Butler first detected a 3.9 Jupiter-mass planet in a 3.3-day orbit around the F8 star Tau Boötis. But whereas Marcy and Butler had used their proven spectroscopy method to make their indirect detection, two years later, a team of British astronomers claimed that they had directly detected the planet's reflected light from the parent star in the mixture of the star's Doppler spectroscopy signal.

Lying more than 15 parsecs away from Earth, the planet orbits its star 20 times closer than Earth does our Sun. In fact, it "roasts away" at an estimated 1,100 degrees Kelvin. The light from Tau Boötis so overwhelms its close companion that even the Hubble Space Telescope

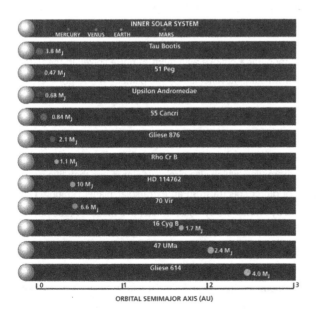

This chart offers a schematic diagram of some of the known Jupiter-mass objects that circle nearby Sun-like stars. (Geoffrey Marcy.)

can't distinguish between the two bodies. Then, in late 1999, after nine nights of observations at the 4.2-meter William Herschel Telescope in the Canary Islands, Andrew Collier Cameron and Keith Horne of the University of St. Andrews in Fife, Scotland, announced that they had found "Doppler-shifted starlight reflected from the planet."[4] Horne went so far as to describe the reflected light as the blue of "faded denim." Understandably, their claim was met with skepticism, for their "direct detection" followed an ill-fated similar attempt by Harvard University's David Charbonneau and Robert Noyes using the 10-meter Keck I telescope in Hawaii. Charbonneau and Noyes wrote afterward that they found no evidence for a highly reflective planet orbiting Tau Boötis.[5] And in their paper published December 16, 1999, in *Nature*, Collier Cameron and Horne wrote that there is a 1 in 20 chance that the "apparent signal is just random noise."

Gordon Walker, who was a referee for the peer review of the U.K. team's paper, said that their direct detection of a planet around Tau Boötis was less a detection than an echo of the primary spectrum. Walker suggested that the actual spectrum should now be even easier to detect in the infrared. But Mark Marley, a planetary theorist at New Mexico State University, acknowledged that if indeed the U.K. team had gotten direct spectra reflected from the planet, the bluish color could best be explained by condensates such as magnesium silicate—basically dirt—in the planet's atmosphere. However, Marley remained skeptical about their claims. "The result is uncertain," he noted. "Pulling such a small signal out of the haystack is an exceptionally difficult process, and there are many ways to go wrong. More and better observations are required before I will believe this result."[6]

Notably, by August 2000, Collier Cameron had publicly retracted his team's claim, following data compilation from newer observations of Tau Boötis that clearly showed that their initial assertions regarding a validated observation of reflected spectra from the planet itself were indeed "spurious." Collier Cameron admitted to me a month after his retraction that there was always a chance that they had simply hit a chance alignment of noise spikes in the data, which mimicked that of a real signal from starlight reflected off the planet. "We've gone as far we can using 4-meter telescopes," he said.[7] To make further headway in obtaining true reflected spectra from the planet circling Tau Boötis, Collier Cameron's team will need time on a bigger telescope or a better spectrograph. They are now in the process of securing both. In the meantime, the British team has turned its sights on obtaining reflected spectra from the innermost planet circling Upsilon Andromedae, which also circles its star in less than five days.

Getting ironclad confirmation of direct spectra from an extra-solar planet orbiting Tau Boötis, a star in a constellation noticed by Homer's Odysseus, would have made for nice poetic symmetry as the old millennium ended and astronomers set their sights on the new. But 20 years after the Canadians began their early Doppler spectroscopy runs, it was clear that the planet-hunters were on track. Pointedly, early in August 2000, William Cochran, an astronomer at the University of Texas in Austin, announced that he and a international team of colleagues had found Earth's nearest extra-solar planet circling Epsilon Eridani, a K2 star that is clearly in what he termed "our own backyard."[8]

When Walker and his colleagues at the University of British Columbia first began Doppler spectroscopy surveys of this star in the late 1970s, they found what he called a lot of "scatter" in the data. Such scatter could account for the perturbation caused by the gravitational tug of an extra-solar planet, yet Walker's data remained inconclusive. Then Cochran and an international team, which included Arnie Hatzes and Barbara McArthur at Austin's McDonald Observatory, successfully combined six different data sets from four different telescopes. They concluded that Epsilon Eridani, which lies only 3.2 parsecs away in the constellation of Eridanus, harbors a Jupiter-mass planet that circles its star once every 6.9 years. With an eccentric orbit that is almost 60 percent that of our own Jupiter, Cochran believes that it is likely that Epsilon Eridani could have room for terrestrial-type planets that may circle the star in shorter orbits. However, the star is so young that it still contains a very visible dust envelope made up of 1-millimeter-sized particles that extend out to 60 AU—more than one and a half times the distance at which Pluto orbits our Sun. The same NASA team that uses dust signatures as a means of detecting evidence of extra-solar planets in orbit around stars with significant dust envelopes, reports circumstantial evidence that Epsilon Eridani may also harbor a planet at least 60 times the mass of Earth. If so, judging from its dust trail, the NASA team concludes that it circles Epsilon Eridani in a very long orbit of 55 to 65 AU. And if this second planet around Epsilon Eridani would be confirmed, it would add another quirk in this ever-expanding *menagerie* of extra-solar planets.

"We're seeing the tip of the iceberg," says Marcy. "Terrestrial planets are the debris of the planet-forming process. An 'earth' is a small fraction of the mass of a protoplanetary disk, so forming an earth is duck soup; it's easy. Look at the Orion Nebula: the stars there are only a million years old, but half of them have protoplanetary disks. That tells me that planets are forming around at least half of all stars. Some might be crummy little Earth-like planets, some of them might be Neptunes or Saturns, but I bet that at least half of all stars have full-fledged planetary systems. I'd bet my house on it."

1   Marcy, Geoffrey, astronomer, University of California at Berkeley. Interviewed on May 25, 1999, at Dana Point, California, and on August 6, 1999, at Hapuna Beach, Hawaii. Follow-ups took place on September 8, 2000, May 10–12, 2001, and June 3, 2001.

2   Mason, Brian D., Project Manager, Washington Double Star Program, U.S. Naval Observatory, Washington, D.C. Private communication with the author on June 11, 2001. Interviewed on July 18, 2001.

3   Apps, Kevin, undergraduate astrophysics student, University of Sussex, U.K. Interviewed on April 12, 2000.

4   Collier Cameron, Andrew et al. "Probable Detection of Starlight Reflected from the Giant Planet Orbiting Tau Boötis." *Nature* 402 (December 16, 1999): 751.

5   Charbonneau, David et al. "An Upper Limit on the Reflected Light from the Planet Orbiting the Star Tau Boötis." *The Astrophysical Journal Letters* 522 (September 10, 1999): L145.

6   Marley, Mark S., planetary theorist, New Mexico State University. Private communication with the author on January 14, 2000.

7   Collier Cameron, Andrew, astronomer, University of St. Andrews, U.K. Interviewed on September 8, 2000.

8   "Search for Extra-Solar Planets Hits Home." McDonald Observatory of the University of Texas at Austin press release (August 7, 2000).

The Thrill of the Hunt

CHAPTER 5

We were sitting at the control console of a brand new $2.5-million telescope, staring at a list of seemingly inauspicious stars. Atop a barren mountain at the southernmost tip of Chile's Atacama desert, there were just the three of us: Michel Burnet, a Geneva Observatory electronic engineer, Hernán Julio, an ESO press liaison, and I, and if we had all collapsed and died in mid-sentence it's certain no one would have come knocking. But isolated as it is, the site is a shrine to technology and the Internet. One of a string of PCs that lined the room was permanently tuned to an observatory weather station, which confirmed what we already knew: it was cloudy, cold, and deadly quiet.

Here at La Silla, surrounded by 825 square kilometers of unpopulated ESO-administered acreage, the Swiss have built a state-of-the-art, fully automated telescope, one of a cluster of 14 optical telescopes and a single radio telescope that can take advantage of about 300 clear nights per year. The newest telescope on site, the one at which I was sitting, officially named the 1.2-meter Leonhard Euler telescope (in honor of Swiss mathematician Leonhard Euler), is owned and operated by Geneva Observatory. It is dedicated to the extra-solar planet-hunting activities of Mayor, Queloz, and their colleagues and is linked directly to the observatory's headquarters via modem, E-mail, phone, and fax.

The arduous route from the coastal city of La Serena to La Silla includes dramatic rocky stretches of barren seacoast, a sharp turn east into a series of steep hairpin grades that criss-cross at least two sets of foothills, and a lengthy descent onto a boulder-strewn moonscape of a plateau. So, after three hours on the road, Julio and I were thinking as much of our stomachs as the stars and luckily pulled in to the observatory just in time for lunch. On any given night, about a hundred staff and observers are on site at La Silla, but most of them never interact, except by chance or in the observatory's well-run canteen. The main administrative building houses a dormitory, a recreation room, and a satellite TV entertainment center the size of a small cinema, which is sometimes also used for film screenings and lectures from visiting astronomers.

The Swiss telescope itself, however, is a world apart. Since its June 1998 opening, its operators have been conducting a systematic survey of 1,562 G, K, and M stars within 60 parsecs of Earth, using their new high-resolution CORALIE spectrograph (which was developed jointly with colleagues at the Observatoire de Haute-Provence and named for the daughter of an OHP engineer). Under good conditions, they average 50 Doppler spectroscopy measurements per night. The astronomical night begins and ends when the Sun is 18 degrees below the horizon. Here, in mid-July, that's roughly equivalent to a 12-hour stint that begins around 7:00 in the evening. The data gathered each night are transmitted directly to Geneva for analysis. So before the astronomers at La Silla are finished for the night, Geneva is already working with the data.

"Usually, I make my list and do a calibration for the thorium lamp," says Burnet, "then I go to dinner; and when I return, I push Start; and if everything is fine I don't do anything until the next morning."[1] Burnet has a general list that he automatically follows over the course of the year, maximizing the time that any given star is in the telescope's field of view. Each observation usually takes 15 minutes per star, plus 2 minutes to set the telescope. Burnet also receives a daily priority list from Geneva. This is the "hot list," so to speak, of stars that may have shown a significant variation in velocity, which could be indicative of the gravitational tug of an extra-solar planet. If, for example, during observations of stars on the general list, one of those stars shows a significant variation in velocity from one observing period to the next, Geneva will normally ask that the star be observed more frequently or placed on the priority list. This enables the observatory to sniff out radial velocity periodicities (sometimes indicative of planets) over a shorter time frame. A software program keeps track of the stars that are new priorities as well as stars from the general observing list. Burnet reviews both to make sure that each night's observing program has an adequate mix from each, and to make sure that the target stars are within the telescope's field of view. But before opening the dome and turning on the telescope, Burnet also has to continually monitor the weather to secure that all of the night's

An exterior view of the dome of the Swiss 1.2-meter Leonhard Euler telescope at La Silla Observatory in Chile. (Photo by Bruce Dorminey.)

targets can be successfully "acquired." There would be no point in wasting half the night tracking stars in one quadrant of the celestial sphere if that part of the sky is lousy with cloud cover. Tonight, for instance, Burnet hasn't even opened the dome, and it's doubtful he will. "There are no optics to look through," explains Burnet. "I spend all night on my chair, and we point the telescope with coordinates. Maybe once a night the system shows us two stars that are so close together [that the computer] asks which one."

So observations sometimes move slowly here, and when the weather causes things to really get sluggish, the scientists often gather in the telescope's fully equipped kitchen for fondue parties or a late-night espresso. Although officially this is not a residence (everyone sleeps during the days down at the main dormitory), the Swiss telescope does have many of the comforts of home: a shower for observers returning from long bike rides around the area, CD players, books, and easy access to the Web. Burnet said that he's known some observers who will open the dome and start the telescope's computers on its observing run and then drive down to the main administrative building to catch a late-night movie. If they are lucky, everything is functioning normally when they return, and there are no angry E-mails from Geneva. Yet in winter, particularly, it's easy to sense that nights alone in this control room could become agonizingly long.

But the team's vigilance has paid off. Aside from their historic 1995 discovery of the planet orbiting 51 Pegasi, the Swiss team, with help from the French in Haute-Provence and new international partners in the U.S. and Israel, has discovered, or co-discovered, more than 30 other extra-solar planets, at least 17 of which have been found on this new telescope. The team is also continuing their long-term survey of 320 G and K stars at Haute-Provence. One of those stars, Gliese 876, a small reddish M star some 4.7 parsecs from Earth, was announced as having a planetary companion. As described at a June 1998 conference in Victoria, British Columbia, the planet is about 1.89 times the size of Jupiter and orbits its parent star every 61 days. The data the Swiss used to detect the planet came from both the Observatoire de Haute-Provence and the new La Silla telescope.

Ironically, at the same conference, Marcy and Butler's team presented data on the same planet from both the Lick and Keck observatories. "Mayor and I both found [the 61-day planet around] Gliese 876 simultaneously and independently," remembers Marcy. "I sat down next to Michel in Victoria and asked, 'I wonder if you have Gliese 876 on your program?' He said, 'I do not know, but will send some E-mails about it this afternoon.' He came back and said, 'You know, we do have the planet on our program, and our velocities show the same planet.' It was the only time in history that a planet was discovered twice at basically the same instant."[2]

Marcy told me all this with a sense of irony, but absolutely no sarcasm. For he knows only too well that once astronomers choose their targets, they often don't pay much attention to the name or number of the target star until it begins to show meaningful perturbations—and even then, among a list of hundreds, one is easy to miss. However, in January 2001, Marcy and his team announced that in addition to the first planet, Gliese 876 b, the parent star, Gliese 876, had one additional planetary companion, called Gliese 876 c, which both Mayor and Marcy had missed in 1998. They missed it because the gravitational perturbation of Gliese 876 c, about half the mass of Jupiter, was masked in the signal of Gliese 876 b, the outer planet. Both planets are in a lockstep, two-to-one gravitational resonance with each other. That means the inner

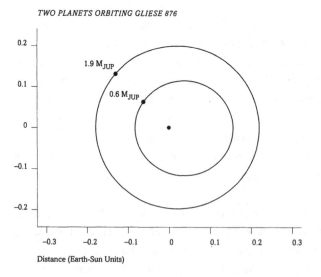

TWO PLANETS ORBITING GLIESE 876

1.9 M$_{JUP}$

0.6 M$_{JUP}$

0.2

0.1

0

−0.1

−0.2

−0.3 −0.2 −0.1 0 0.1 0.2 0.3

Distance (Earth-Sun Units)

The orbits of the two known planets circling the star Gliese 876. (Geoffrey Marcy, U. C. Berkeley.)

planet makes two journeys around the parent star for every one journey that the other planet makes around its parent star. Theorists are still trying to understand the significance of such resonance, as it is the first time the phenomenon has been detected in an extra-solar planetary system. They think it might be caused by one planet having moved from its original location, resulting in both planets gravitationally pushing and pulling, or "shepherding," the other in order to maintain this synchrony.

EXPANSION TEAMS

Encouraged by the phenomenal success of the teams led by Marcy and by Mayor, other groups of astronomers have begun developing their own planet-hunting alliances. Martin Kürster, a longtime ESO astronomer who spends 105 nights per year at La Silla, recently led an international team that found a 2.3-Jupiter-mass planet circling Iota Hor, a naked-eye G star in the southern constellation of Horologium (Pendulum Clock), 17 parsecs from Earth.

Kürster's team used ESO's 1.4-meter Coude Auxiliary Telescope, also at La Silla, to make the discovery. According to Kürster, the planet circling Iota Hor has a slightly eccentric 320-day orbit, which means that if imposed on our Solar System, it would cross both Venus and Earth orbits in elongated fashion. Moreover, based on five and half years of observations from a 40-star survey begun in 1992, Kürster doesn't rule out the possibility that Iota Hor could also harbor other planets.[3] This find is encouraging because it shows that discoveries are possible even when surveying a limited number of stars on a very modest telescope. It's also an indication that this burgeoning planet-hunting effort has taken on a grassroots appeal among observational astronomers the world over. Astronomers who only ten years ago would have eschewed the idea of using valuable telescope time to look for extra-solar planets are being drawn into this ongoing quest.

Marcy's team, meanwhile, has been guaranteed ten years' observing time at Keck, where they are surveying 450 G, K, and M stars, most of which are within 50 parsecs of Earth. One of their goals is to detect a Jupiter-like planet—a Jupiter analog—in a 12-year orbit (the actual Jupiter orbits the Sun in 11.862 years). "We've already got three years in the bag," says Marcy, "so, ten years from now we will have enough data to look for planets in 12-year orbits. Are there analogs or clones of our own Jupiter? We don't know. We haven't found any true twins of Jupiter yet."

At California's Lick Observatory, Marcy's team has been surveying 107 stars since 1987, and they've added 200 more under the supervision of Butler and Debra Fischer, Marcy's postdoctoral researcher. "Within 10 parsecs there are dozens of M dwarfs," says Marcy. "M dwarfs are the most numerous low-mass type of star. It's like grains of sand on the beach—there aren't too many rocks on the beach, but there are a lot of tiny grains of crystal."

But in time, Jupiter anologs will be found, and when they are, their discovery will give credence to the notion that, around the stars surveyed thus far, we've only been able to detect the bizarre 4 percent of extra-solar planets—the ones that are in eccentric, or oddly elliptical, and very short orbits around their parent stars. We know that our own Jupiter was crucial to the formation of our stable Solar System. Therefore theorists reason that if we are to find extra-solar planetary systems that resemble the Solar System, we must first find extra-solar Jupiter-like planets which circle their own Sun-like stars roughly every 12 years. (This will be discussed in detail in Chapter 6).

However, with planet hunting still in its earliest stages, we must be patient. Before astronomers can find the full range of planets that must exist out there, they must first hone the requisite planet-hunting technology.

ADVANCING THE HUNT THROUGH DOPPLER SPECTROSCOPY

To date, Doppler spectroscopy searches have been subject to a skewed view of extra-solar planetary companions. Thus far, all the planets discovered with this technique have been massive gaseous-type planets, most in unusually short orbits around their stars. (To be absolutely sure that they've found a planet above the noise in the data, Doppler spectroscopy surveyors must observe a planet's gravitational perturbations over two orbits around its parent star.) As a fledgling technology, Doppler spectroscopy surveys were sensitive only to giant Jupiter-mass planets with orbital periods that allowed the observers to validate their data over short time scales. Today, both teams have made great strides in improving their accuracies, allowing them to find lower-mass planets at greater distances from their parent stars. However, both teams still have the same ultimate goal: to achieve accuracies approaching 1 meter per second. At this writing, the best either team can claim is 2 meters per second, and more typically they reach accuracies of 3 to 5 meters per second.

Mayor says that his team is working on High Accuracy Radial Velocity Planetary Search (HARPS), a joint project with ESO that would result in the installation of a new spectrograph at La Silla's 3.6-meter telescope, the largest on the mountain. "Basically, it's the same ELODIE-type instrument that we have at Haute-Provence," he states, "but with much higher resolution."

Meanwhile, at Haute-Provence, François Bouchy, an astrophysicist in the service of France's Conseil National de Recherches Scientifiques (CNRS), has been slaving over a very complex laser contraption that he claims will totally take the guesswork out of Doppler spectroscopy measurements. He and two colleagues have taken over the second floor of the observatory's 1.52-meter telescope. Essentially, Bouchy would like to use lasers in tandem with a conventional xenon gas spectrograph to eliminate the effects of Earth's own changes in velocity, thereby subtracting them from the overall equation of Earth-based Doppler spectroscopy. At the time of my visit in May 1999, however, they were facing a multitude of challenges—not the least of which was a broken prism and budget problems. If they should be successful, Bouchy believes that his team's Absolute Astronomical Accelerometer (AAA) technique would eventually be even more accurate than anything Marcy or Mayor are planning.

Nevertheless, says Bouchy, "All the teams are limited by noise. We will try to reach the photon noise limit with a little telescope before going to a bigger telescope. With the 1.52-meter here, we estimate that we can reach 1 meter per second. The difficulty is that when you observe the star, you measure the radial velocity of the star and the radial velocity of Earth in its revolution around the Sun. In one year, the velocity of Earth is plus or minus 30 kilometers per second."[4] In order words, sometimes it's aperture, not accuracy, that counts. While Bouchy and his colleagues are experimenting on a small aperture telescope to sort out the vagaries in their radial velocity accelerometers, they are ultimately hindered by that telescope's "photon noise limit," which is a way of saying simply that the telescope has reached the edge of its capability to accurately delineate light from any given target. Beyond that point, the angular resolution of the star becomes enmeshed in noise, or muck, caused by the limitations of the telescope's optics. While accuracy is very important, at times the aperture size also is significant. As an example, in early April 2000, Marcy's team announced that using Hawaii's 10-meter Keck I, they had detected two Saturn-mass extra-solar planets circling two different stars. (Jupiter alone is three times the mass of Saturn.) One, a planet at least 70 percent the mass of Saturn, lies in a 76-day orbit around 79 Ceti, a G5 star 36 parsecs away in Cetus, the Sea Monster. The other, at least 80 percent the mass of Saturn, is in a three-day orbit from HD 46375, a K star 33 parsecs away from Earth in the Monoceros constellation.

"We won't find anything lower [in mass] than Saturns," says Marcy. "That's about it. We might find a few planets that are smaller, but mostly we are stuck finding Saturns and Jupiters, and maybe Neptunes if they are very close in and they jerk violently on the star. But I bite my lip when I say that because we are now in an era where the expectations are watched way in advance of what we can actually do."

PLANETARY EUREKA

Back at La Silla, the Swiss dome was still closed. Burnet was being polite, but I sensed that we had upset his normal routine—so much so that with an air of resignation and perhaps against his better judgment, he allowed me to look at his list of targets, which included notations by the Geneva astronomers. As I clicked through the list, I found something interesting and pestered him with yet another question. There were two priorities on his list, one of which he couldn't observe even if the conditions had been right, because it wasn't in La Silla's field of view at that time of year, and another one that was outside the astronomical night. One of the priorities, HD 130322, was flagged with the notation "planète connue" (known planet), and was also listed as having a very short periodicity like that exhibited by a 51 Pegasi-type "hot Jupiter." The star was categorized as stable and as having no significant fluctuations—a good sign that its changes in radial velocity were in fact being caused by a planet and not noise from a chromospherically active star. (Doppler spectroscopy astronomers usually try to avoid stars that appear to be chromospherically active, because their outer layers can cause radial velocity fluctuations, which can sometimes be misconstrued as mimicking the actions of a planet on its parent star.) But this particular star—HD 130322, from the Henry Draper catalog—was also listed as TOP, top priority. Burnet told me that even if he was finally able to measure this particular star before the night was over, it was unlikely that Geneva would list it for more observations for at least a few weeks. As he pointed out, over a period of a year, it's not always necessary that a star be measured continuously. Burnet referred to a diagram on the PC monitor, in which tiny dots can represent individual star measurements over weeks, days, months, and sometimes years. He explained that the object of the ongoing observations is to look for a pattern, so that over time what is known as a "light curve" (or, precise observational record of a given star) can be drawn between the individual dots, with each dot representing a single measurement. A double-point on the chart, for example, simply means that the telescope has taken two different measurements of the same star on the same night.

Burnet explained that the interest in HD 130322 stemmed from the fact that (as was clear from looking at the diagram) there did seem to be a periodicity, or pattern, in the star's measurements, potentially indicating the presence of an unseen companion. Burnet noted that if Geneva was convinced that the periodicity in evidence on the diagram was due only to chromospheric activity (noise), then by now the search team would have taken the star off its observing list. Yet here it was, still listed as a high priority. Unfortunately, the observers were only halfway there. Although there were obviously patterns in the variations of HD 130322's velocity, the team wasn't ready to publicly announce that the variations were caused by a planet.

Michel Burnet of Geneva Observatory at the control console of the Swiss 1.2-meter Leonhard Euler telescope at La Silla Observatory in Chile. (Photo by Bruce Dorminey.)

An overview of the European Southern Observatory's extensive La Silla facilities. (European Southern Observatory.)

With a last flick of the mouse, our time with Burnet had come to an end. The Sun was 67 degrees below the horizon. Dewpoint was −10, dry enough for a camel. But with clouds everywhere and no wind, it didn't look good for observing. Burnet said he would wait until three o'clock, and if the weather didn't break, he would head back to the dormitory. Julio and I adjourned to the canteen for an early morning steak.

Less than two months later, the Swiss made it official: HD 130322, a K star in Virgo, does have at least one Jupiter-mass companion that circles around it every 10.7 days. And by August 2000, the Swiss had announced that the 1.2-meter telescope had also found what at that time was only the second known extra-solar planetary system circling a Sun-like star. It consists of two planets circling HD 83443, a Sun-like K star only 80 percent of the mass of the Sun, located some 43 parsecs away from Earth in the constellation of Vela. The inner planet has a Saturn-like mass and circles the star almost every three days—at that time, the shortest orbit of any known extra-solar planet. The second (or outer) planet is about half Saturn's mass and circles its star almost every 30 days. The Swiss report that, incredibly, despite eccentric orbits, this bizarre Saturnian-like system appears to be dynamically stable (able to maintain its equilibrium even as its planets continue on very odd orbits around the parent star.) That's relatively easy in a planetary system like our own, in which—with the exception of Pluto—all nine planets orbit their star in almost circular orbits. Otherwise, the gravitational weight from one planet could perturb another to such an extent that one would eject the other out of its orbit altogether. So, at first glance, two Saturn-mass planets forming a system in very close eccentric orbits around a star such as HD 83443 would logically seem to lend itself to instability. But that was not the case here.

Perhaps the real news is that a little more than 20 years after Gordon Walker pioneered Doppler spectroscopy as a method of planet detection, the plot has thickened. In fact, the pace of discoveries has become frenetic; with each new discovery of smaller and smaller-mass planets, astronomers are giving planetary theorists new data with which they can craft their ever-changing theories of how planets form, evolve, and mature into full-fledged systems.

1   Burnet, Michel, electronic engineer, Geneva Observatory. Interviewed on July 12, 1999, at La Silla, Chile.

2   Marcy, Geoffrey, astronomer, University of California at Berkeley. Interviewed on May 25, 1999, at Dana Point, California, and on August 6, 1999, at Hapuna Beach, Hawaii. Follow-ups took place on September 8, 2000, May 10–12, 2001, and June 3, 2001.

3   Kürster, Martin, astronomer, ESO La Silla, Chile. Interviewed on August 5, 1999, at Bioastronomy 99, Hawaii.

4   Bouchy, François, astrophysicist, Conseil National de Recherches Scientifiques (CNRS). Interviewed on May 6, 1999, at Observatoire de Haute-Provence, France.

CHAPTER 6

The Lithium Test

If Marcy's dream of finding planetary systems around half of all stars is based in reality, finding a real Jupiter analog would be an encouraging first sign that such systems may indeed be common in our galaxy. To comprehend the potential impact of finding such extra-solar Jupiter analogs, it is first necessary to assess the impact that our own Jupiter has had on the stable evolution of our Solar System. The mass of Jupiter is greater than the mass of all of the Solar System's other planets combined. At 5 AU, or five times the distance between Earth and the Sun, it stands sentry to the outer Solar System.

More than 4 billion years ago, Jupiter's role was that of a planetary garbage collector, swallowing up or sweeping the inner Solar System free of marauding planetesimals. In this role, Jupiter offered gravitational insurance that the habitable zone in which Earth was forming would enable the inner planets to develop stable, circular orbits. Without Jupiter acting as what Marcy terms the "bully" on the block, Earth would have been slammed around to such an extent that it would soon have seen the backside of our Sun.

Millions of such ejected "Earths" are thought to be wandering aimlessly throughout the interstellar medium. "Jupiter," says Marcy, "dictated that the only orbits that would be stable in our Solar System are nearly circular ones. Circular orbits were the only survivors; the others were goners. If Earth were not in a circular orbit, it would have been ejected, and we wouldn't be here talking about it. Within a couple of years, we are going to find 'Jupiters' at 5 AU. If these planets are in eccentric orbits, then our Solar System will be rendered unusual in a fundamental way. I don't think our Solar System is one in a million, but it's one in a thousand."[1]

THE REAL THING

Just what is a real "Jupiter"? Our Jupiter is a gas giant with a density 1.33 times that of water, and large enough to hold 1,300 Earths. With its swirling, complicated atmosphere and brightly colored clouds, Jupiter looks as exotic as they come. In composition, it is 93 percent hydrogen; the rest is helium, nitrogen, oxygen, and carbon, in the form of ammonia, methane, and water. NASA's Galileo space probe also found that Jupiter has trace amounts of neon, argon, krypton, and xenon in its atmosphere.

Jupiter's core is thought to be metallic hydrogen, perhaps rocky, perhaps not. According to William Nellis, a physicist at the Lawrence Livermore National Laboratory in California, Jupiter's core must be gravitationally sufficient to hold hydrogen molecules. In conventional accretion theory, once Jupiter has enough initial mass in the form of a rock and ice core at least ten times the mass of Earth, it can begin accreting (or holding) hydrogen gas.[2]

Jupiter was able to form its icy core where it did, at 5 AU, because that's the distance at which water ice was first stable. Closer in to the Sun, temperatures were high enough for water ice to vaporize. So, even though it was possible for rocky material to accumulate, there wasn't

Jupiter as seen in this composite photo by NASA's Cassini spacecraft on December 7, 2000. (NASA/JPL/University of Arizona.)

enough additional mass (which in Jupiter's case was provided by ice) to add to the rocky core. When Jupiter and the other gaseous giants—Saturn, Neptune and Uranus—formed, ice served as planetary "glue," enabling the accumulation of enough rock and additional ice to capture (or gravitationally bind) hydrogen gas from the nascent solar nebula to the planets' cores. In Jupiter, eventually, the sheer mass and density of this accretion process compressed molecular hydrogen into liquid hydrogen, and then finally metallic hydrogen.

Until 1996, metallic hydrogen had never been created in the laboratory; that's when Nellis and his colleagues succeeded in doing so using a two-stage hydrogen-propelled gas gun fired upon a target of liquid hydrogen (hydrogen cooled to 200 degrees Kelvin, or −100 degrees Fahrenheit). The metallization takes place at pressures equal to 1.4 million times that of Earth, and at temperatures of 30,000 degrees Kelvin (53,540 degrees Fahrenheit). Because Jupiter's mass is 318 times that of Earth, we might never know for certain whether Jupiter has a rock and ice core of 10 Earth masses. But University of Arizona models based on data collected by NASA's Galileo spacecraft, which has been surveying Jupiter since 1995, suggest that the planet's gravity can be explained without a rock and ice core. If so, then Jupiter may have formed from a cloud of molecular gas, in a process known as gravitational disk instability. In contrast to the conventional accretion mechanism involving rock and ice cores, which uses the gravity of the core to attract and accrete hydrogen gas in the disk instability formation process, these giant protoplanets would independently contract into spherical bodies from large clumps of gas and dust. This process would not be unlike the way our own Sun formed, or the Milky Way collapsed from clouds of molecular gas and dust. Except, this planet formation process would take place inside the disk that makes up a protoplanetary nebula surrounding a young star. As a process of planet formation, it would be quite separate from the previously mentioned planetary accretion process taking place elsewhere in the protoplanetary nebula.

First proposed in 1951 by the Dutch-American astronomer Gerard Kuiper, this 50-year old theory of gravitational disk fragmentation was given new impetus some five years ago by Alan Boss, a planetary theorist at the Carnegie Institution. At the moment, the debate still rages over which theory offers the most likely scenario for giant planet formation. The answer may come in the near future, as the result of a proposed $300-million NASA mission, called the Interior Structure and Internal Dynamical Evolution of Jupiter (INSIDE Jupiter) mission. INSIDE Jupiter would consist of a spacecraft orbiting the giant planet in a very close polar orbit to conduct extensive mapping of Jupiter's gravitational and magnetic fields. The spacecraft, which should see launch no later than 2006, would also look for irregularities and perturbations in Jupiter's magnetic and gravitational fields, which would indicate whether Jupiter has a metal core. But even as astronomers learn more about the exact origin and makeup of Jupiter, they will also continue looking for its analogs around other stars.[3]

"The holy grail," says Boss, "is going to be finding a real Jupiter with a Jupiter mass and a Jupiter orbit that leaves enough room inside for a terrestrial planet to have a stable orbit and be habitable. We've gained a lot of confidence because we know these wild, wacky, hot 'Jupiters' exist. We didn't even know that they existed five years ago."[4]

KEEPING APACE OF CHANGE

In the summer of 1999, extra-solar planet hunters and planetary theorists descended upon Flagstaff's Northern Arizona University for a weeklong conference. There, they debated the boundaries of planet and stellar formation, for the new extra-solar giant planets (EGPs) are making theorists scramble for new explanations.

It would be hard to overstate the pace of change that is taking place in this burgeoning field of planet hunting. In the process of researching and writing this book, my own point of view on what constitutes a planet has changed almost as many times as those of the theorists. When I began my initial research, people I met while traveling would often ask me to explain the difference between a star and a planet. I always gave them a confident, even smug, answer: "A star is a burning ball of nuclear fusion," I'd begin. "Planets are the bodies that orbit them." But the more I've learned, the more I realize how little that definition is really worth.

To comprehend the theoretical gap between planets and stars, one must first come to at least a rudimentary understanding of so-called brown dwarfs. According to a standard definition, brown dwarfs are stellarlike objects that range between 13 and 75 Jupiter masses (Mj). They are formed as stars, by the gravitational collapse of a fragment of molecular hydrogen, typically about a tenth of a parsec across. The distinction between a Jupiter analog and an ordinary brown dwarf can be made by their mass. Depending on its metallicity, any stellar object above 75 masses of Jupiter will begin full-blown hydrogen burning. As University of Arizona astronomer Adam Burrows describes it, this is defined as the limit above which 100 percent of the energy losses from the star's surface are compensated by thermonuclear burning in the center. Brown dwarfs by contrast will burn deuterium only, a primordial hydrogen isotope and relic of the early Universe.[5]

Brown dwarfs can burn deuterium for as long as a billion years, then they cool off, making them even more difficult to spot. In Hollywood terms, these are starlets, not stars. But they don't go unnoticed. More than 100 brown dwarfs have been found free floating and around other stars.

WHAT'S IN A NAME
The term "brown dwarf" was coined in 1975 by Jill Tarter, who, at the time, was a doctoral candidate in astrophysics at the University of California at Berkeley. In search of a good thesis topic, part of her research involved determining whether brown dwarfs (then still theoretical objects) might make up the majority of the galaxy's dark matter. Technology at the time precluded her getting a good model of what such objects would look like in the color realm, so, as Tarter explains, "I finally just decided to call them brown, because brown is not really a color." (Brown is perceived by the human eye as a mixture of red, green and blue wavelengths of light and is therefore not considered a primary color.)[6]

Making the distinction between a gaseous giant planet and a brown dwarf is only a matter of semantics, according to Bill Hubbard, planetary scientist at the University of Arizona and a participant in the 1999 Flagstaff conference. In terms of their interior physics, Hubbard believes that they are identical: "If you were to take a Jupiter-mass chunk of the solar atmosphere and cool it down to Jupiter temperatures, then you would see the same soup of molecules as in Jupiter. Jupiter and Saturn are two hydrogen-rich planets and are the closest analogs in our Solar System to the exo-planets that we are studying. It's always been my claim that Jupiter is indistinguishable from a brown dwarf. The same physics apply and the same theory applies."[7] Brown dwarfs are generally identified by lithium lines in their spectra. And because lithium is among the first chemical species to be destroyed when a star starts burning hydrogen, when astronomers spot lithium in a substellar object, it generally means that the object is cool enough to avoid hydrogen-burning. In the classic definition, brown dwarfs are hot enough to burn a limited supply of deuterium, but not massive enough to start life as a full-fledged hydrogen-burning star. Main-sequence stars tend to have burned through their supply of lithium early in their careers as hydrogen-burners, while gaseous giant planets are cool enough so that lithium tends to bind with other chemicals to make compounds that don't show up as pure lithium in spectroscopy. That's true in almost all the brown dwarfs except Gliese 229B, first spotted in late 1994 at Mount Palomar Observatory in California and confirmed in 1995 by the Hubble Space Telescope.

At roughly 25 Jupiter masses and an age of 1 billion years, Gliese 229B orbits Gliese 229, an M star, at a distance of 45 AU, and lies some 5.8 parsecs away in the Lepus constellation. While Gliese 229B shows no signs of lithium, it does show methane, which is only visible in objects cooler than 1,000 degrees Kelvin (or about three times hotter than the highest setting on a conventional kitchen oven). By contrast, Jupiter is very cool; its upper atmosphere tracks a characteristic radiating temperature of 124 degrees Kelvin.

Desert Wanderers

Even after the discovery of Gliese 229B, brown dwarfs were thought to be rare. Until late 1999, only four *bona fide* brown dwarfs were known to orbit other stars. Marcy had even coined the term "brown dwarf desert" to explain the absence of objects in the brown dwarf-mass range. But by the turn of the millennium, the brown dwarf desert began to look a lot less barren. Astronomers making sensitive, wide-field surveys of very faint objects in nearby star-forming regions were reporting hundreds of extraordinarily faint low-mass brown dwarf-type objects. But unlike Gliese 229B, these low-mass objects were so-called free-floaters, apparently wandering aimlessly among dozens of other similar companions in young star clusters.

One such cluster is visible to the naked eye and lies in the vast star-forming region of the Orion Nebula, which can be made out as the middle "point of light" in the sword that forms part of the Orion constellation. In April 2001, two U.K. astrophysicists, Philip Lucas at the University of Hertfordshire and Patrick Roche at Oxford University, reported their analyses of infrared spectra obtained from 13 faint points of light in the Orion Nebula. Their spectra had been taken during 1999, using the 3.8-meter U.K. Infrared Telescope atop Mauna Kea in Hawaii. While conducting their analyses, they found three of these objects to be at or below the deuterium burning limit, at 8, 11, and 13 Mj.

"There are probably on the order of a billion of these objects floating around in our own galaxy," says Roche. "There is a kind of continuous distribution, but my best guess is that they formed through fragmentation of clouds like stars, instead of like planets. If they formed from the remnants of a stellar disk [in the manner of planets], they would be moving away from the stars in which they were formed."[8] As the gap between giant gas planets and low-mass stars begins to fill, the lines between planet, brown dwarf, and low-mass star become increasingly skewed. And with each new discovery, theorists rush to come up with new ideas about how such low-mass objects could form by conventional stellar fragmentation processes.

In January 2001, at a meeting of the American Astronomical Society in San Diego, Marcy, Butler, and their colleagues announced that they had discovered a pair of bizarre planets in orbit around HD 168443. This star, roughly 38 parsecs away from Earth in the constellation Serpens, is circled every 58 days by a 7.7-Mj planet, a distance about a third closer than Mercury's orbit of the Sun. A second object orbits the star every 4.8 years, or more than twice as long as it takes our Mars to orbit the Sun. But unlike Mars, the second object in this system is a real whopper, better characterized as a brown dwarf in orbit around a star than a planet. It is the most massive extra-solar "planet" ever found, and as Alan Boss wrote in the January 25,

This composite mosaic image, taken with the European Southern Observatory's Very Large Telescope (VLT) in Chile, shows the central part of the star-forming region of the Orion Nebula (M42) with the four main stars that make up the so-called Trapezium *near the center. (European Southern Observatory.)*

2001, issue of *Nature*, together, both of HD 168443's planet-like companions contain "at least 25 times the mass of the planets in our Solar System."[9]

The most striking question is, how did this outer object form in such close proximity to its parent star? Boss explains in the article, "There are two competing mechanisms for the formation of gas giants in a protoplanetary disk: core accretion and disk instability. Core accretion could account for the formation of planets with masses up to about 5 Mj, but whether it could produce objects as massive as the companions of HD 168443 within the lifetime of a typical protoplanetary disk (a few million years) remains to be seen." Gilles Chabrier, an astrophysicist at France's Ecole Normale Supérieure in Lyon sees no point in even attempting to answer this question at this stage: "You can spend days arguing about whether to call Marcy's 17-Mj object a planet or a brown dwarf, but that's a waste of time. It's going to take a decade before we have a reasonable understanding of how these extra-solar planets form. People are rushing to judgment. We need more studies, so we have to be patient."[10]

EYE ON THE PRIZE

Marcy explains that when his group of astronomers first began their surveys, they expected to find an equal mixture of planets and brown dwarfs. Instead, they found very few brown dwarfs. Mayor reports similar results: some of the brown dwarf discoveries that he announced in his initial Doppler spectroscopy surveys have been upgraded to very low-mass M stars. So at least for now, brown dwarfs as companions to normal Sun-like stars appear to be rare. The real goal of the planet hunters remains finding a planet with roughly the mass of Jupiter in a Jupiter-like orbit from its star. Marcy believes that in time at least some of those stars may be found to harbor Jupiter analogs at distances of 5 AU, which would still leave the door open for the forma-

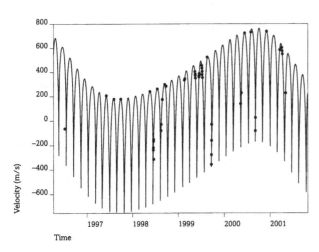

The radial velocity curve of the star HD 168443, which lies some 38 parsecs away from Earth in the constellation Serpens. The periodicities found in its radial velocity curve indicate that the star is orbited by at least two massive bodies, one of which lies well above the conventional mass range of planet. (Geoffrey Marcy, U. C. Berkeley.)

tion of Earth-like planets at distances closer in to their parent star. For theorists now believe that close-in Jupiters are anathema to the formation of a planetary system with Earth-mass planets in stable inner orbits.

As in all branches of science, a contrarian has emerged. David Black, a theoretical astrophysicist and planetary scientist at the Lunar and Planetary Institute in Houston, Texas—and a longtime proponent of planet hunting around nearby stars—agrees that the planet hunters' data are real, but he disagrees with how that data have been interpreted. In a paper he presented at the Flagstaff conference, Black notes, "It is more likely that the low-mass companions form a homogeneous class, with their underlying set of orbital and mass distributions being essentially indistinguishable from those of stellar binary companions."[11] Black contends that both astronomers and theorists are too eager to conclude that just because an object has a minimum of less than 10 Mj, it must automatically be dubbed "planet." He calls such definitions "scientifically useless."

Black's opinion hasn't swayed Didier Queloz, Mayor's co-discoverer of 51 Pegasi, who was also at the conference in Flagstaff. He flatly stated that he was going to call anything below the 13-Mj deuterium-burning limit a planet. To which Black replied: "I think these low-mass companions are different from brown dwarfs, but they're not planets. They could be a new class of objects. The challenge is [to answer]: How do you make them? The easiest explanation for me is that they form like stars. But how the hell does nature make such low-mass objects via a process that is similar, perhaps identical, to the way it makes binary stars? If you assume they are planets, then you've got to come up with a number of *ad hoc* mechanisms, such as orbital migration [whereby giant gas planets form at relatively large orbital distances from their star and then are pulled into closer stellar orbits], to explain how these objects came from far away from the star to within a few stellar radii. What is often overlooked is that orbital migration produces circular orbits, not the eccentric orbits that all of these companions have."

A few weeks after the conference in Flagstaff, I caught up with Alan Boss, who had worked under Black as a postdoctoral researcher. Boss counters Black's assertions by pointing to the data on Upsilon Andromedae and its three Jupiter-like companions, which are in relatively close orbit around the star, noting that no one has ever seen a binary system like that. Black remains unconvinced. He doubts that the Upsilon Andromedae system could maintain its long-term stability. He is also suspect of the interpretation of the data regarding its proposed middle companion. The middle "planet" has an orbital period of 241 days, which, Black says, isn't consistent with a well-constrained and stable planetary system. "If there are three companions," says Black, "then it looks very much like a planetary system. If [there are] two companions, with the middle period in the data being due to something other than a companion, then you have a very tight binary and a wide, eccentric-orbit, triple-star system. Multiple star systems look exactly like that."

As for the most recent detection of Saturn-mass objects by Marcy's group, Black argues that brown dwarfs in the Orion cluster have been found with as little as 8 Mj. "That would suggest," says Black, "that nature knows how to make such low-mass companions by a star-like process, even if we do not. Most of the companions with short periods, ten days or less, have probably experienced substantial mass loss. So if the true mass is as low as that of Saturn, we are probably just seeing the stripped core of what was a larger object. And that's totally different from our planetary system, the companions to a pulsar, and to the patterns that we see in the regular satellite systems of the giant planets." He concludes, "Unfortunately, the people who should be the most critical in these discussions are the theorists. I've been very disappointed in my colleagues in that regard. I'm not saying that there aren't planetary systems out there—I think they *are* out there—but I think they are probably rarer than most people suspect."

Black's best guess is that the new extra-solar planets have evolved parallel to, and perhaps in a very similar manner as, traditional binary star systems. But again, the dynamics and formation of binary star systems are also not well understood.

"Everyone is entitled to their opinion," says Marcy. "I have the data, and we have all the planets. The objects we are finding are all piled up at 1 to 3 Jupiter masses. Black doesn't respect the data. He has this idea that they are not planets, and then he does everything he can to support his point of view. That's not science. There are 20,000 practicing astrophysicists on this planet, and precisely two are resistant to the idea that these are planets. One of them is David Black. When the data are in conflict with a scientist's theories, then the scientist has two choices, to reject either the theory or the data. Black rejects the data."

1   Marcy, Geoffrey, astronomer, University of California at Berkeley. Interviewed on May 25, 1999, at Dana Point, California, and on August 6, 1999, at Hapuna Beach, Hawaii. Follow-ups took place on September 8, 2000, May 10–12, 2001, and June 3, 2001.

2   Nellis, William J., physicist, Lawrence Livermore National Laboratory, Livermore, California. Interviewed on April 19, 2000.

3   Bergstralh, Jay, Cassini program scientist, NASA headquarters, Washington, D.C. Interviewed on March 29, 2001.

4   Boss, Alan P., planetary scientist at Carnegie Institution, Washington, D.C. Interviewed on August 7, 1999, at Bioastronomy 99, Hawaii.

5   Burrows, Adam, planetary theorist and astronomer, University of Arizona at Tucson. Interviewed on June 11, 1999, at Flagstaff, Arizona.

6   Tarter, Jill C., director of SETI Research, SETI Institute. Interviewed on August 5, 1999, at Bioastronomy 99, Hawaii.

7   Hubbard, Bill, astronomer and planetary scientist, University of Arizona. Interviewed on June 12, 1999, at Flagstaff, Arizona.

8   Roche, Patrick, astrophysicist, Oxford University. Interviewed on March 29, 2001.

9   Boss, Alan P. "Giant Giants or Dwarf Dwarfs." *Nature* 409, no. 6819 (January 25, 2001): 462.

10   Chabrier, Gilles, theoretical astrophysicist, Ecole Normale Supérieure de Lyon, France. Interviewed on March 29, 2001.

11   Black, David C., theoretical astrophysicist and planetary scientist at Lunar and Planetary Institute, Houston, Texas. Interviewed on June 10, 1999, at Flagstaff, Arizona. A follow-up took place on April 19, 2000.

      Black, David C. and Tomasz Stepinski. "A Statistical Assessment of the Assertion That Low-Mass Companions to Stars Are Extra-Solar Planets." Abstract presented at the From Giant Planets to Cool Stars conference, Flagstaff, Arizona, June 8–11, 1999.

CHAPTER 7

Expanding the Search

When Einstein first advocated the notion that gravity could bend light, space, and time, there seemed to be little chance that he or anyone else would be able to verify it. Certainly, no one would have guessed that eventually the theory would be used to jump-start galactic surveys for extra-solar planets. Einstein had predicted that the Sun should gravitationally displace, or bend, incoming starlight by as much as 1.75 arcseconds—a significant amount (roughly equivalent to the separation of two car headlights as seen from a distance of 200 kilometers). In fact, it was only three years after he made the concept a central tenet of his 1916 theory of general relativity that a total eclipse of the Sun offered a way to test the theory. In May 1919, Arthur Stanley Eddington, an English astronomer, organized two teams of British astronomers—sending one to Brazil and the other to the West African island of Principe—to look for signs that the Sun was gravitationally bending light from the Hyades star cluster.

Eddington went with the Principe team. Although the African island was plagued by mediocre weather, he was able to take photographs (a feat in itself, as photography was then a largely new technology), showing the relative position of the Hyades cluster the night before the eclipse. More remarkably, during the eclipse, he made 16 shots of the Sun. After several months of analysis, Eddington was able to announce that, indeed, there had been a shift in the stars' apparent positions due to the eclipse, and that they roughly matched what Einstein had predicted. As a result, Einstein, then 40 years old, became an international celebrity. By the end of the century, ESA's Hipparcos satellite had proven Einstein's theory correct to within one part in a thousand.

Einstein's theory, however, went further. He had also postulated that given the right alignment among stars along our own line of sight, a foreground star can gravitationally "lens," or bend and brighten, light from a more distant background star, causing a ring-like image to form. The effect, known as microlensing, is a natural manifestation of the relativity theory, by which the foreground star, generally less luminous than the background star, acts as a gravitational lens, magnifying the background star's light over a period of several hours or weeks. Generally, as seen from Earth, the background star is about 8,000 parsecs away, while the star acting as the lens lies closer, at a distance of some 6,000 parsecs. In such a scenario, it is impossible to identify either star, as the foreground star becomes distorted into a ring of light surrounding a natural lens. In other words, the foreground star is being gravitationally probed by light from the background star. While the foreground star acting as the microlens cannot actually be identified in a star catalog, microlens observations are still useful for theorists taking statistics in searches for dark matter or unseen matter that might make up the missing mass of the Universe. These observations are also useful for planet hunters who are trying to set parameters on the numbers of extra-solar planets. They can use the background star's light as a searchlight that, essentially, sparks through the foreground star's lens, thus enabling them to detect planetary companions. Ironically, Einstein never believed microlensing would prove to be a practical astronomical tool. For once, time has proven him wrong.[1]

MONITORING MICROLENSING EVENTS

Since the first confirmation of a microlensing event in 1993, groups now routinely monitor the center of our galaxy for such phenomena, which are best seen from April to November in the Southern Hemisphere. At that time, Earth is aligned to see the foreground disk (or spiral structure) of the galaxy against its more distant central galactic bulge (central hub of the Milky Way galaxy). The chances of such precise alignments are literally one in a million, which is why monitoring groups look at very crowded fields of stars at least 1 degree across. To understand what they're up against, imagine looking at the monitor of a typical desktop PC, with an on-screen image of stars as thick as snow. At La Silla Observatory, Amy Stutz, an Ohio State University physics major, spent part of one 1999 semester doing an observing stint for the French-run Expérience pour la Recherche d'Objets Sombres (EROS), an ongoing gravitational lensing and microlensing search, at ESO's 1-meter Marly telescope. "There is probably a microlensing event in there now," says Stutz, pointing at the star-filled PC monitor in front of her, "yet just by looking at the screen, it's not apparent. What we are looking for is an increase in the brightness of the star versus the previous day's image. We are doing a light curve, so it's luminosity against time. On a good night, we get over a 100 images, which are simultaneously processed through red and blue filters for two separate images—that's to avoid variable stars."[2]

When EROS or other microlensing monitoring groups find events, they post them on the Internet. But when I walked into the control room at the Yale CTIO 1-meter telescope atop Cerro Tololo mountain in July 1999, Bruce Atwood was using the Internet to check baseball scores. This meant only one thing: it was cloudy yet again. Atwood, along with Andy Gould and Darren DePoy, all astronomers and members of an Ohio State University team hunting planets by microlensing, were the principal brains behind a new-fangled contraption called A Novel Double-Imaging CAMera (ANDICAM) that allows for simultaneous optical and infrared imaging. Funded in part by the U.S. National Science Foundation (NSF), it took three years to build. It's about the size of a good-sized tree trunk and processes stellar images to the very high precision necessary to follow up on microlensing events. The camera literally takes photons of light collected from the same telescopic target and simultaneously separates them into visible and infrared wavelengths, allowing the astronomers to observe the celestial target in two different wavelengths at the same instant. Although this is not necessarily helpful for all observations, it is particularly useful when tracking ongoing microlensing events, as this simultaneous imaging over a wide wavelength spectrum gives the astronomers more precision in determining whether planets are indeed circling the microlensing event's foreground star.

The Ohio State University team poses with ANDICAM, their double imaging infrared/visible light camera on the ground floor of the Yale 1-meter telescope at CTIO in Chile. From left to right: Darren DePoy, Bruce Atwood, Jerry Mason, and Ken Sills. (Photo by Bruce Dorminey.)

The Ohio State team is part of the Probing Lensing Anomalies NETwork (PLANET), a six-year old international group of planet hunters. PLANET members, using a network of telescopes in South Africa, Tasmania, Western Australia, and Chile, contribute their time and effort to work as one observing team, on which it can be said that the sun never rises. They follow hundreds of microlensing events over several weeks, measuring them roughly every hour. Such events typically involve a fainter lower-mass F, G, or K star acting as a lens for a brighter and generally higher-mass background star. Thus, the stars that are probed naturally via microlensing are those that are more likely to have planetary systems.

"If a star suddenly gets fainter and brighter very quickly, we know that it has a companion," says PLANET team member Scott Gaudi, a former Ohio State University astronomer, now at the Institute for Advanced Study at Princeton, New Jersey. "The signature of a planet in a microlensing event is a short deviation spike. This is the signature that there is a planet around the lens doing the microlensing. The duration of the spike tells us the mass ratio between the primary lens and the companion lens. We can guess what the mass of the planetary companion is by the time scale of its spike relative to the time scale of the main microlensing event itself."[3]

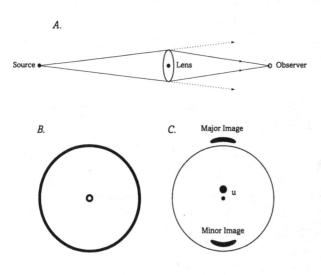

A.

Source ● Lens Observer

B. *C.* Major Image

o

u

Minor Image

BASIC SINGLE LENS MICROLENSING

Panel (A) shows a side view of the geometry of microlensing. Light emanating from the source is bent by the gravitational field of the lens toward the observer. The actual path of the light rays is shown in a solid line. The path the light rays would travel in the absence of the lens is shown in a dotted line.

Panel (B)—the observer's point of view as seen when the observer, lens, and source are perfectly aligned.

If the lens is not perfectly aligned with the source, as in panel (C), then there are two images of the source, a major image and a minor image. (Scott Gaudi.)

A Jupiter-mass planet will generally create a spike in the microlens of a foreground star that will last for a day or two. The PLANET group asserts that their telescopic instruments are attuned to identify planets that would be orbiting the foreground star at distances of up to 6 or 7 AU. Because of this, planet hunters using the microlensing method find their research very complementary to Doppler spectroscopy searches. In normal Doppler spectroscopy searches, the teams' instruments are generally most sensitive to planets with high masses and in short orbits around their star. By contrast, microlenses are sensitive to 1-Mj planets that circle their stars in even longer orbits than our Jupiter. But these are for single-lens events. Binary stars have a different signature and are tricky. They can even be confused with a planetary signature.

The Sign of Two

The Microlensing Planet Search (MPS), a group at the University of Notre Dame, made head-lines in 1999 when they reported that, after two years of data analysis, they had seen a plane-tary signature in the lens of a binary star system. Although the PLANET team agreed that the lensing event MACHO-97-BLG-41 (first detected by a group looking for dark matter and sub-sequently monitored by MPS) did not have the characteristic signature of a single lens, the two teams disagreed on the reasons why.

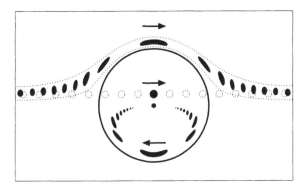

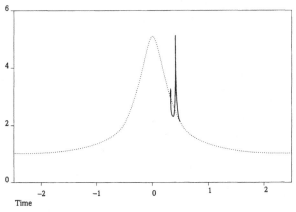

BASIC PLANETARY MICROLENSING

Top panel: As the source moves from left to right, the major image also moves from left to right, while the minor image moves from right to left. The primary lens is shown as a black dot. If the primary lens has a planet in the path of one of the images, i.e., within the black dotted lines, then the gravity of the planet will fur-ther bend the light rays from the source, creating a perturbation on the single lens light curve.

Bottom panel: The single lens light curve is plotted against time and magnification. Here a pertur-bation caused by the presence of a planet in orbit around the fore-ground star (the star acting as the microlens) is modeled as an amplitude spike, which protrudes from the single lens curve. (Scott Gaudi.)

WHAT'S IN A NAME
MACHO-97-BLG-41 is an astronomical notation to signify the following: the year 1997's forty-first
gravitational-lensing, or galactic bulge (BLG-41), event detected by the MACHO Project, a U.S.-led
international group of astronomers, who at the time were looking for MACHOs (Massive Compact
Halo Objects). MACHOs are a theoretical form of dark matter thought to exist in the halo of our
galaxy. (Our galactic halo is roughly spherical in shape and extends out from the galactic center.)

After analyzing its data, MPS announced the discovery of a planet of about 3 Mj orbiting a binary system at a distance of 7 AU, while the two stars were only 1.8 AU apart. In other words, MPS believed that it had detected two stars in very close proximity to one another—less than twice the distance from Earth to our own Sun—being orbited by a planet three times the mass of Jupiter. Again, however, data interpretations give rise to disagreements. The Ohio State team (a member of the PLANET network) is strongly convinced that the Notre Dame team (MPS) simply "misdiagnosed" a light curve of a very short-orbit binary star system involved in a complicated process of microlensing. "We definitely see the right number of binary stars that we would expect, which is a good indication that we are doing things right," says Darren DePoy.[4] At the 2000 annual meeting of the American Astronomical Society held in Atlanta, Scott Gaudi presented two years' worth of PLANET data. He noted that PLANET, after five years of operation, had yet to detect planets via microlensing. Gaudi also reported that, based on their most recent data, less than a third of the G and K star populations they had been monitoring could harbor a real "Jupiter."[5]

"In the next ten years we are going to be able to put real statistical limits on the frequency of planet formation around other stars," promises DePoy. "That's one of the things I wish the radial velocity people would do instead of saying, 'Oops, we found another one; oops, we found another one.' I don't like to speculate about things. . . . The interesting question is, How many planets are out there? And this is a quick way to get an answer to that question." Incredibly, even after their spectacular Atlanta announcement, a review board at the U.S. National Science Foundation balked when the Ohio State team asked for more funding. DePoy said one member of the NSF board dismissed their work as not "important enough." "Right now," DePoy counters, "we can make the statement that less than a third of all G and K stars we are studying have Jupiter-mass planets between 1 and 5 AU. There are papers that go back five years which state unless a Jupiter-mass planet forms 5 AU from its star, it's impossible for an Earth-mass planet to form. If the condition for having an Earth-like planet around a star is that you have to have a Jupiter-like planet farther out, then there aren't very many stars that have terrestrial planets. At the end of ten years, we will have a good idea of what the number is, not just an upper limit. I think we're going to find that no more than 10 percent of stars like the Sun are going to have planetary systems that resemble our own, and the number could be a lot less than 10 percent. I don't actually find studying planets that interesting—the weather on Jupiter is less interesting to me than the weather in Columbus, Ohio. But I do think that finding out how many stars out there have 'Jupiters' or planets like Earth is a very profound question."

PLANET DETECTION VIA OCCULTATION

Detecting planets in the gravitational microlens of a foreground star requires extreme precision and care, particularly when it comes to data analysis. Likewise, detecting Jupiter- and even Earth-mass planets as they transit across the face of a Sun-like star requires equal amounts of patience and precision. But unlike finding planets in a microlens, which is a one-time, one-shot event, detecting planets via occultation often enables ongoing verification that the planet indeed is there. Thus, skeptics can sometimes simply be swayed by a trip to the telescope. This was the idea that occurred to Geoffrey Marcy on November 6, 1999, while going through preliminary analyses of observations at the Keck I telescope in Hawaii. It was clear from the data that there was a planet about half the mass of Jupiter in a very short 3.524-day orbit around HD 209458, a G0 star some 47 parsecs away from Earth in the Pegasus constellation. Marcy was determined to quiet the skeptics with a double whammy of proof, using two methods of planet detection on this star almost simultaneously. So, that same day, Marcy E-mailed Greg Henry, an astronomer at Tennessee State University in Nashville, who had been working informally with Marcy for over a year trying to detect transits of planets in short orbit around their stars. Until that point, none of Henry's observations for transits had produced positive results.

OCCULTATION
Occultation refers to an astronomical body that is passing in front of another, thus obscuring its view. Asteroids, planets, and natural satellites can all obstruct the view of other celestial bodies, which is usually evidenced by the fact that light from the body being obstructed dims wholly or in part. Strictly speaking, occultation can occur in any part of the electronic spectrum, although it is generally used to describe events in the visible spectrum involving the Moon or one of the planets of our Solar System obscuring background stars.

"The message I got from Marcy was just one or two lines of text," says Henry, "giving me the name of the star, the planet's period, and that it was a good one to search for transits. I looked at the data and realized that the next possible transit was that same weekend, on Sunday night."[6] Although Henry lives and works in Nashville, thanks to high technology and the Internet, he observes using a robotic (automatic) 80-centimeter telescope at Fairborn Observatory, a one-man operation in Arizona's Patagonia Mountains on the old silver-mining frontier just shy of the Mexican border. Upon Marcy's E-mail, Henry simply prepared an observation request that contained the target star's coordinates, its immediate stellar neighbors, and the optimum sequence of observations. He uploaded the information in the form of a large ASCII file directly to the telescope. The telescope then automatically carried out the observations overnight, while Henry got a good night's sleep in Nashville. The results were waiting for him in an FTP file when he arrived at work Monday morning. The data confirmed that indeed there was a planet in a very short orbit around HD 209458 by showing a 1.6-percent dip in the star's luminosity over a three-hour period due to the transit of the newly detected planet.

This photo shows the automatic telescopes of Fairborn Observatory in the Patagonia mountains of southern Arizona. The 0.8-meter telescope in the foreground is one of the automated telescopes used by Tennessee State University astronomer Gregory Henry to make extremely precise measurements of brightness changes in Sun-like stars and to search for extra-solar planets. With this telescope, Henry discovered planetary transits in the star HD 209458, which resulted in direct measurements of its planet's mass, radius, and density. (Photo courtesy Gregory W. Henry, Tennessee State University.)

Once astronomers know the spectral type and approximate age of the parent star—in HD 209458's case, a G0 star slightly larger than but roughly the same age as our Sun—they are able to extrapolate details about the transiting planet's mass, density, and size. (The planet was also independently detected via Doppler spectroscopy, and subsequently confirmed via occultation by David Charbonneau, Timothy Brown, and their colleagues at the High Altitude Observatory in Denver. Then, in April 2000, the Hubble Space Telescope confirmed that this planet could be detected regularly transiting its star.)

"There are about 16 short-period planets," Henry explains, "and [with the exception of HD 209458] I have surveyed them all with negative results. If you look at the statistics, the odds of finding a short-period planet at random is about 10 percent." That's because the planet is in such a short orbit; there is only a 1 in 10 chance that it will be seen edge-on in relation to Earth's own line of sight. As it happened, HD 209458 was the tenth star that Henry surveyed.

During that same November in 1999, two astronomers in California were continuing their collaboration to find Earth-like planets in occultation transits around nearby M stars. Laurance Doyle from the SETI Institute in Mountain View and Lee Rottler at Lick Observatory and the University of California at Santa Cruz had spent a thousand hours doing a pilot survey of CM Draconis, an eclipsing binary system composed of two M4 stars 16.5 parsecs away from Earth. Using a 90-centimeter telescope at Lick Observatory, they have reported a successful trial run of their transit timing method, which uses a complicated mathematical algorithm in tandem with ongoing observations of the star to match the best window of opportunity to view an Earth-mass planet. But to date, no such planet has been found, and the only planet that has been detected transiting its star remains HD 209458.

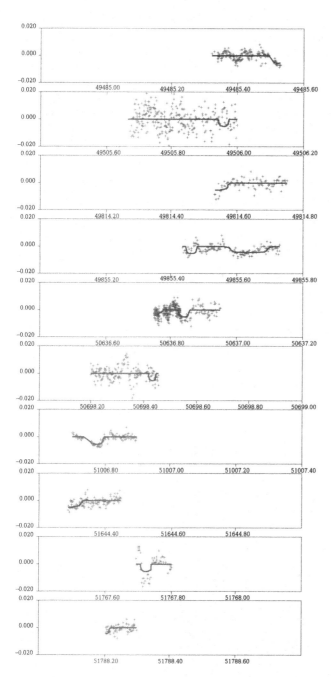

Light change of a 20.56-day candidate planet transit across the two components of the CM Draconis eclipsing binary star system. The solid line is the best model fit to the brightness changes of CM Draconis compared with standard stars (differential magnitude). Time on the horizontal axis is in Heliocentric Julian Day, i.e., the time in the center of the Sun in number of days since noon on January 1, 4713 B.C. (Model calculations by Hans-Jörg Deeg.)

ECLIPSING BINARIES

In binaries (double stars), eclipsing refers to the fact that in the course of their orbits around each other, one star eclipses (or occludes) the other either partially or wholly on a periodic basis. As Doyle explains, because such binaries are known to eclipse the other at periodic intervals, it follows that eclipsing binaries are "edge-on" (or very nearly "edge-on") to our line of sight. Thus, if one or both of the stars harbor planets, these planets would also be more likely to move through a so-called transit window, in which an orbiting planet is subject to detection by measuring dimming effects caused by the planet's transits of the eclipsing binary system.

"What we are doing is purposely looking at low-mass M binary stars," says Doyle. "Across this M binary, a Neptune would cause a 1-percent decrease in luminosity, and Earth would cause about a 0.01 [percent] decrease. Eclipsing binary stars are already edge-on [or in the same orbital plane as viewed from Earth], so we have a better chance of getting a transit. Our main interest at SETI is to show that detection of terrestrial-class planets is possible right now. It's like hitting people over the head to convince them of this, but that's what we are trying to show from the ground right now, [with] no more technology needed."[7]

Doyle and Rottler have also surveyed stars in the galactic plane for transits of Jupiter-like planets. Rottler used data collected on CTIO's 4-meter Blanco telescope in Chile to analyze light curves from stars in the globular cluster NGC 6752 in the Pavo (Peacock) constellation. Best seen from the Southern Hemisphere, this cluster is estimated to be at least 12 billion years old. Doyle and Rottler predict a success rate of 38 percent for transits in its core region, which is located along the plane of the Milky Way. They are also analyzing data collected on 103 of 108 eclipsing binaries in Baade's Third Window, part of a very young crowded field of stars that

Image of Baade's Third Window in the galactic plane, taken with the 0.9-meter telescope at the Cerro Tololo Inter-American Observatory in 1998. (Photo by Laurance R. Doyle.)

is also in the galactic plane. "Before we can apply the transit detection algorithm," says Rottler, "we need light curves, and getting those light curves on stars in a crowded field is not a trivial task. But if we do detect planets in the binaries in the older globular cluster, it may tell us something about long-term stability of planetary systems. If we don't find planets, or find relatively few planets in the older cluster, relative to the younger [field of stars], it may tell us that most planets get destroyed, or 'eaten,' by their stars. Or it could mean that tidal interactions with other stars rip them up. In the next few years, we'll build up statistics in both populations of stars, then we may be able to constrain some of the theories of planetary evolution."[8]

Their task will not be trivial, as members of a Hubble team found out in 1999. During eight days of observations in July of that year, Ron Gilliland, an astronomer at the Space Telescope Science Institute in Baltimore, led a team of astronomers using the Hubble Space Telescope to target 47 Tucanae, a very old cluster of G- and K-type stars in the southern constellation of Tucana. Using Hubble's Wide Field and Planetary Camera 2 (WFPC2), Gilliland targeted some 35,000 stars (out of a possible million that make up the cluster) in a field of view that spanned almost 3 parsecs. The team looked for dips in stellar luminosity that would signal Jupiter-mass planets in three- to five-day orbits around parent stars in the cluster, which includes stars up to 12 billion years old, some of them 4,600 parsecs away from Earth.

Gilliland calculated that, based on the ages of the stars' masses, their likely inclinations to Earth's line of sight, and Hubble's own sensitivity limitations, if such Jupiter-mass planets were there, then at least 17 such planets would be identified *en masse*. But the team came up empty-handed—possibly because very old clusters may not be a good place to look for planets, as

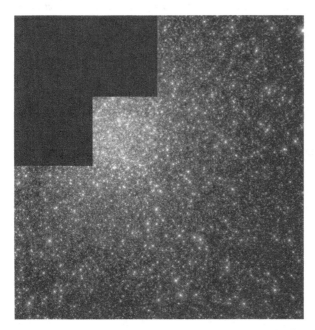

Over a period of eight days, the Hubble Space Telescope observed a "swarm" of 35,000 stars in 47 Tucanae, a globular cluster located in the southern constellation Tucana. Astronomers expected to find 17 extra-solar planets. Instead, they found none. This could indicate that conditions for planet formation and evolution may be different in different regions of our Milky Way galaxy. (NASA and Ronald Gilliland, Space Telescope Science Institute.)

they may lack the metallicities to form them. In other words, they lack carbon, iron, oxygen, and other so-called stellar metals, precisely because they are so very old and haven't gone through the recycling processes needed to form heavy metals, and thus, planets.[9]

Two dedicated space-based photometers are also in the works to expand this ever-changing planet-hunting effort. The first, COROT (for Stellar COnvection and ROTation), is a French spacecraft now scheduled to be launched in 2004 by the French Space Agency. It will use a 27-centimeter telescope to observe at least 60,000 stars, broken into five target fields of 12,000 stars each. COROT will look for hot Jupiters closer than 0.3 AU to their parent stars. And it is anticipated that it may find 13 to 22 Jupiter-mass planets with orbits of less than 50 days. ESA is also studying a $170-million Earth-orbiting spacecraft dedicated to photometry. Dubbed Eddington, in tribute to Arthur Stanley Eddington, the aforementioned English astronomer, the project is now listed as a reserve mission, with no set launch date. However, if it does launch, it would use its 1-meter telescope to conduct high-precision, space-based, wide-field photometry on hundreds of thousands of stars at once. It would look for Earth-mass planets making transits, which would be evidenced by a dimming effect equal to only about 1 part in 10,000 as the planet passes across the face of its parent star.

PHOTOMETRY

Photometry can almost be considered a science unto itself. Like spectroscopy, it has evolved to have a major day-to-day impact on astronomy in general. In its more basic form, photometry simply allows for the measurements of a star's brightness and color, which in turn enables astronomers to make assumptions about a star's internal evolution and processes. Photometers, instruments that use filters to make precision measurements of light coming from any given star or field of stars, are particularly useful in measuring the dimming effect caused by a planet transiting its parent star. Photometry is also used by astrophysicists interested in asteroseismology, the measurement of Doppler shifting of global oscillations taking place on the surface of other stars. Asteroseismology is crucial in helping theorists hone models of stellar spectral types.

PLANETARY DETECTION VIA MASERS

Cristiano Cosmovici, a Rome-based cometary physicist with Italy's Institute for Space Science, may have happened upon what is arguably, to date, the most imaginative way to look for extrasolar planets. He spent the summer of 1994 watching Jupiter for the impact effects of Comet Shoemaker-Levy. And while Jupiter's atmosphere was being splayed by more than 20 cometary nuclei, Cosmovici and his colleagues at Italy's Medicina 32-meter radio telescope at the Instituto Radio Astronomia in Bologna, serendipitously detected a water maser line emitting in the 22-gigahertz radio spectrum. Once the comet hit the upper atmosphere of Jupiter, the cometary fragments of Shoemaker-Levy exploded, and the water inside burst forth and expanded in the planet's upper ionosphere. That, in turn, excited the water molecules into a maser, which was detected by Cosmovici. While he had proposed looking for water from the comet during its bombardment of Jupiter, he never would have guessed that in the process he would observe the first water maser ever detected in our Solar System. Unfortunately, such cometary impacts on Jupiter occur only once every million years or so, meaning that neither Cosmovici nor any of his colleagues will be around to confirm the observations.

Undaunted, Cosmovici plans to pursue this basic concept elsewhere, namely in other planetary systems. He reasons that if extra-solar Jupiter-mass planets are bombarded by comets, then, in principle, he could observe the same effect. For now, Cosmovici is using the 32-meter radio telescope five days a week to target ten F, G, and M stars in an effort to locate bombardment of an extra-solar "Jupiter," or simply the natural water-masing from an extra-solar Earth-like planet. Such water-masing can also occur on terrestrial-type planets that orbit stars with high infrared emissions. (Earth's own atmosphere has never been a water-masing candidate simply because its chemical makeup doesn't allow for it.) "First you see a signal," says Cosmovici. "You don't know if it's an Earth-like planet or a cometary bombardment on a giant planet. If, after a period of observations, this signal disappears, it means it was a bombardment. But a signal that remains means it is coming from the atmosphere of an Earth-like planet. Then you have to make a statistical model to see which Doppler shift would practically align with a planet in order to determine the orbital period of the planet, and thus determine the distance from the star."[10]

Cosmovici applies the same methods in the radio spectrum that Doppler spectroscopy searches use in the optical. In his observations, however, Cosmovici looks for Doppler velocity variations of the stars over an 8-megahertz bandwidth. Currently, he is only targeting stars out to 15 parsecs away from Earth; but after the new 64-meter radio telescope in Sardinia is completed in 2003, Cosmovici's observing runs will be made a hundredfold more efficient, and allow an increase in both distances and targets.

Cosmovici would probably acknowledge, privately if not publicly, that such methods are a long shot. But then, who would have ever dreamed that the first extra-solar planets ever discovered would orbit a pulsar, and that they would be found using a radio telescope? Although Doppler spectroscopy will likely continue to be the main tool for planet hunting for the next three to five years, astronomers are becoming more and more technologically ingenious in learning to draw water from a stone. Until they have grand space-based telescopes, which can truly detect and image extra-solar planets *en masse*, their burgeoning efforts at ground-based planet hunting can use all the ingenuity the field can muster. For, as each new bizarre discovery reminds us, we remain a long way from being able to characterize the exact nature and frequency of extra-solar planetary systems that must exist even within our local neighborhood.

1   Guillermier, Pierre and Serge Koutchmy 1999. *Total Eclipses: Science, Observations, Myths and Legends*. London: Springer-Verlag Praxis Series: 102.

2   Stutz, Amy, Ohio State University undergraduate physics major. Interviewed on July 14, 1999, at ESO's 1-meter Marly telescope at La Silla, Chile.

3   Gaudi, B. Scott, astronomer, Institute for Advanced Study, Princeton, New Jersey. Interviewed on June 10, 1999, at Flagstaff, Arizona.

4   DePoy, Darren, astronomer, Ohio State University. Interviewed on July 11, 1999, at the Yale CTIO 1-meter telescope, Chile.

5   Gaudi, B. Scott and colleagues at the PLANET Collaboration. "Microlensing Constraints on the Frequency of Jupiter Mass Planets." Abstract presented at the American Astronomical Society Annual Meeting, Atlanta, Georgia, January 2000.

6   Henry, Gregory W., astronomer, Tennessee State University, Nashville. Interviewed on April 21, 2000. A follow-up took place on June 8, 2001.

7   Doyle, Laurance R., astronomer, SETI institute. Interviewed on August 7, 1999, at Bioastronomy 99, Hawaii.

8   Rottler, Lee, astronomer at Lick Observatory, California. Interviewed on June 8, 1999, at Bioastronomy 99, Hawaii.

9   Gilliland, Ron, astronomer, Space Telescope Science Institute, Baltimore, Maryland. Interviewed on April 5, 2001.

10   Cosmovici, Cristiano, cometary physicist at Italy's Institute for Space Science (IFSI), Center for National Research, Rome. Interviewed on August 6, 1999, at Bioastronomy 99, Hawaii.

CHAPTER 8
Planetary Demarcations

However infrequent, however primitive, life's first criterion is that a habitable body lie at a habitable distance from its star. No matter who's ultimately right—Black (who believes that we only have proof of planets around pulsars), Marcy (who bets there are "full-fledged" planetary systems around half of all Sun-like stars), or DePoy (who ventures that no more than 10 percent of Sun-like stars have planetary systems resembling our own)—the mere fact of finding planets around other Sun-like stars only marks the beginning of a decades-long quest to find planets that could ultimately harbor life. And depending on the temperature and type of star, the habitable zone can vary wildly.

There are no hard-and-fast rules, but theorists believe that habitable planets are most likely to form around stars that "live their lives" in the stellar slow lane. The more massive the star, the faster it burns, hence the quicker its demise. Stars with a hydrogen-burning life of more than 2 billion years make the best candidates for creating planets within a habitable zone. These include F, G, K, and M stars. Because all stars evolve along their main-sequence (or hydrogen-burning) phase, the habitable zone expands outward as the star ages. At the time of Earth's formation, our own G2 spectral-type Sun had a radiation output some 30 percent less than it has today. So, over the ages, the boundaries of our Solar System's habitable zone have expanded outward as well. At the moment, the Solar System's habitable zone is estimated to lie between 0.95 and 1.37 AU in a relatively tight band comprising Earth's current orbit.

At 5,770 degrees Kelvin, our Sun is heating up by approximately 1 percent every 100 million years. This slow evolution has very little effect on global warming, or on the Sun's 11-year solar cycles, when periods of high activity (caused by solar flares, sunspots, and other emissions) can wreak havoc on everything from agriculture to telecommunications. Such cycles may indeed have an impact on the year-to-year vagaries of our global climate, but they do not figure in the long equation of our planet's climatic stability in the same way that our Sun's evolution has in the past and will continue to do in the future. "What we think limits the inner edge of the habitable zone," says James Kasting, a planetary scientist at Penn State University, "is what I've termed a 'moist greenhouse atmosphere.' That's where you still have liquid water on the surface, but you get enough water in the atmosphere that you get a wet stratosphere. Once that happens, photodissociation [molecular breakup as caused by photon radiation] of the water begins—when hydrogen escapes into space and the oxygen goes back and reacts with the planet's crust, and you lose the oceans. In our simple models, we would place the inner edge of Earth's habitable zone today at about 0.95 AU (or, 5 percent closer to the Sun). We think that if you push the Earth out you would tend to build up carbon dioxide gas (CO_2) in its atmosphere. There's a feedback mechanism whereby CO_2 tends to accumulate if the surface temperature gets colder. But there's a limit to that, because at some point, the CO_2 begins to condense and form CO_2 clouds, and CO_2 ice deposits at the poles. So the outer edge of the habitable zone is determined by this condensation of CO_2. That edge is negotiable."[1]

THE CLIMATIC BALANCING ACT

Earth's climate requires a careful balancing act involving silicate weathering, a continuous recycling mechanism that enables the recycling of carbon from the atmosphere into Earth's crust and then back again. The process is estimated to be sufficient to recycle all the carbon in both Earth's atmosphere and its oceans once every 400 million years. The carbon is literally "rained out" in Earth's weather systems. Then, it chemically bonds with the Earth's crust where, in a process known as subduction (where one crustal plate is buried underneath another), the CO_2 returns to Earth's tectonic plate system as carbonate. There, it is exposed to very high temperatures and pressures and is spewed forth once again as CO_2 through volcanic activity or through the ocean's many hydrothermal vents. This self-regulating mechanism is one key to Earth's capability to hold and maintain a stable atmosphere and climate.

In general, the larger the planet, the more capable it is of "holding on" to its atmosphere. That's just gravity. Larger planets also have internal sources of heat, which create the tectonic activity necessary for such recycling. (Tectonic activity of some sort is essential for any terrestrial planet to reprocess its carbonate rocks and return its CO_2 to the atmosphere.) Mars never had what it took to do so. Venus had the atmosphere, but lost what water it did have over a billion years ago. With it, Venus also lost its capability to recycle its atmosphere in the same way that Earth has done throughout the ages. Although Earth does lose some hydrogen and helium into the relative vacuum of space, Mars lost most of its atmosphere some 2 billion years ago, at about the same time it died geologically.

Life on Mars? That Is the Question

So the debate still rages over whether there is, or ever has been, microscopic surface or subsurface life on Mars. Some planetary scientists believe crude bacteria-like microbes could still exist there, even if they are buried under several meters of Martian permafrost. If it ever existed above ground, such microbes would have to have been capable of withstanding extreme changes in temperature. But we do know that at one time, the red planet was host to significant amounts of liquid water on and beneath its surface. During its orbit around Mars in 2000, NASA's Mars Global Surveyor spacecraft returned images of gully-like surface formations, indicating that the red planet did have flowing liquid water at some point during what the space agency has termed its "recent geological past." However, "recent geological past" could mean anything from last week to a few million years ago. Life might also be present in a netherworld, for example, beneath one of the frozen oceans of Europa, one of Jupiter's largest moons. But even if they do harbor simple life forms, neither Mars nor Europa are exactly gardens of Eden.

LUNAR SERENDIPITY

The real cause of Earth's relatively stable climate is more likely the result of a catastrophic chance encounter "only" 100 million years after its formation (4.5 billion years ago). Planetary theorists call it a hard capture, an astronomical euphemism for the fact that Earth got nailed at high speed by a Mars-sized rock. The impact literally ripped out a large chunk of Earth's hide and sent it into Earth orbit. It also stirred up a vapor cloud of silicate dust, which eventually accreted into our Moon.

In early 2000, planetary scientists at the Southwest Research Institute in Boulder, Colorado, announced that computer simulations they had run supported evidence of this hard capture by tracing back the Moon's orbits over the eons to the time of its formation. Unlike the Solar System's more than 70 known natural satellites, our Moon is now tilted by roughly 5 degrees in relation to Earth's own orbit around the Sun. By contrast, satellites circling other planets in our Solar System only show an orbital tilt of 1 to 2 degrees. The Southwest researchers believe that the Moon's odd tilt is an orbital smoking gun that points to a giant impact from a Mars-sized rock at least 4.4 billion years ago. This is the age of the oldest lunar rocks collected during the Apollo program. William Ward and Robin Canup, two Southwest

This composite triptych of our Moon was made from three images taken by NASA's Cassini spacecraft. The image serves to visually reinforce what we often forget—that at over a quarter Earth's diameter, the Moon is large enough that some planetary scientists have termed our Earth-Moon system a double planet. (NASA/JPL/Caltech/Cassini Imaging Team/University of Arizona.)

Institute planetary scientists, used numerical models in computer simulations to trace the orbital parameters of the Moon from its inception to its present 5-degree tilt, where it now lies some 400,000 kilometers from Earth. Based on their models, they report that shortly after its initial formation, the primordial Moon must have had an orbital tilt of 10 to 15 degrees.

Ward and Canup started with the assumption that at the time of its formation, the Moon initially had only a 1-degree orbital tilt. They calculated that, within a year's time, the Moon initially coalesced at a distance of only 14,000 miles on the outer edge of the debris disk, which had been created by the displacement of material following the impact. "It may have taken 100 million years for the lunar crust to cool and solidify," says Canup, "but we've got a Moon that has just formed with this leftover inner debris disk. Over hundreds of years, the gravitational interaction of the Moon with the disk generated waves in the Moon's debris disk that gave the Moon its 10- to 15-degree tilt."[2] These first hundreds of years also coincide with the time it took for the Moon's debris disk to dissipate. Then, over billions of years, the Moon eventually reached its present-day expanded Earth orbit.

Regardless of its tilt or random formation, our Moon is proportionally so large in relation to Earth that it is markedly different from the Solar System's other satellites. The Moon is a quarter of Earth's diameter and 1/81 of Earth's mass. While Jupiter, Saturn, and Neptune have satellites of comparable mass, each of those planets is many times more massive than Earth.

OBLIQUE VARIATIONS

In 120 B.C., the Greek astronomer Hipparchus noticed that Earth's rotational axis varies. It doesn't always point toward Polaris, or Alpha Ursae Minoris, an F star 250 parsecs away (which in its present position can be seen at less than 1 degree from the celestial north pole). Over a 26,000-year period, Polaris marks out a large circle on the sky as Earth precesses—essentially,

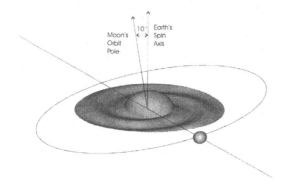

This computer model done by planetary scientists at the Southwest Research Institute simulates the evolution of the Moon's unusual 5-degree orbital tilt. This tilt is further evidence that the Moon was formed from a chance cataclysmic impact from a Mars-sized body. (Ward & Canup 2000.)

wobbles like a spinning top—around the Sun. Known as the *precession of the equinoxes*, this effect is due to gravitational forces exerted on Earth's equatorial bulge by the Moon and the Sun. (The equatorial bulge is an increase in the equatorial diameter caused by the centrifugal forces of Earth's rapid rotation.) The Moon accounts for two-thirds of these precessional forces, while the Sun makes up the remaining third.

Centuries after Hipparchus' observation, Urbain Jean Joseph Leverrier, the French astronomer who discovered Neptune in 1846, was the first to figure out the long-term variations in Earth's obliquity. Nearly a century later, in 1930, Serbian astronomer Milutin Milankovitch wondered if this change in obliquity was the root cause of the ice ages that were set off by variations in terrestrial insolation, the Sun's radiation as it strikes the Earth's atmosphere at the polar latitudes.

If not for the Moon, Earth's obliquity (or axial tilt) would be wildly chaotic, varying by as much as 60 percent over a period of 2 million years. And at certain points in time, Earth would reach obliquities of up to 90 percent. The poles would melt, resulting in utter climatic chaos. Mars, in contrast, has two small moons and an obliquity that varies from 0 to 60 degrees, making it impossible to predict its climate over millions of years. Earth's obliquity currently varies by only 1.3 degrees. This variation is, however, totally unrelated to normal seasonal changes caused by Earth's inclination of 23 degrees 27 minutes. According to Jacques Laskar, a celestial mechanic at the Observatoire de Paris who has done years of extensive calculations on the effects of the Moon on Earth's obliquity, this "little shaking of the orbit of our Earth is just enough to add a small oscillation of Earth's rotation axis. This oscillation causes a small ±1.3-degree change in Earth's obliquity, which is the source of the change of climate during the ice age, known as the *Milankovitch cycle*."[3]

Laskar contends that the Moon's serendipitous influence on Earth's obliquity means that the likelihood of finding extra-solar planets with stable obliquities within habitable zones is probably less likely than winning the lottery. "The main point is that we are not in a generic situation," states Laskar. "You had to have this random catastrophic impact that formed the Moon. I'm not a biologist, but we would not be here if it weren't for the Moon."

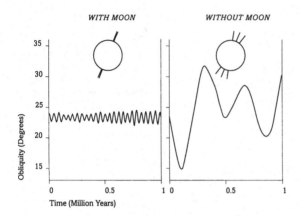

In this graphic, the stabilizing influence of the Moon is readily apparent. Without the Moon, the changes in Earth's obliquities would become clearly evident over time scales of millions of years. (Jacques Laskar/CNRS.)

Yet Kasting and his colleague Darren Williams, also a planetary scientist at Penn State, contend that all is not lost in the gambit to find habitable, stable climates around Sun-like stars, even if a moon the size of ours is a rarity for an Earth-like planet. The Penn State researchers used an energy-balance model to simulate Earth's climate at obliquities of up to 90 degrees. They of course acknowledge that at such a tilt, Earth's climate would become what they term "regionally severe," with large seasonal cycles and temperature extremes, but they point out that other Earth-like planets might fare differently, depending on their land–sea distribution, as well as their position in the habitable zone.[4]

Kasting and Williams also point out that the farther away a planet is from its star, the less it will be affected by obliquity changes, because it can respond by accumulating CO_2 in its atmosphere that would self-regulate its climate as a counter to such extremes. Finally, they conclude that even with high obliquities, a "significant portion" of extra-solar "earths" might still be habitable. Even though these are points well taken, another Penn State researcher, meteorologist Gregory Jenkins, modeled Earth during the Archean period. This epoch began some 3.8 billion years ago and featured only a slightly elevated carbon dioxide level. It was a period when Earth was tilted on its axis by 70 degrees due to the recent impact that led to the creation of the Moon. Jenkins' model also factored in Earth's geography as it stood roughly 3.8 billion to 2.5 billion years ago. He concluded that the Earth's surface was 95 percent ocean at the time.[5]

Jenkins believes the oceans were the key to maintaining Earth's warmth even at that angle. He also found evidence for three different epochs of global glaciation over the last 2.5 billion years. This made him ask how nascent life on Earth could have survived. To answer that question, he ran another model, with more land mass but at the same 70-degree tilt. He found—as Kasting and Williams asserted—that even though the equators are glacial at such angles, the poles remain unfrozen. In and of itself, this fact would mean some species could have survived.

Although today Earth remains tilted by some 23 degrees, Jenkins figures that over millions of years, it has gradually come untilted from its 70-degree angle, most likely due to a massive buildup of ancient continents at the South Pole, near the end of the Precambrian era more than 570 million years ago. Over a 100-million-year period, the influence of their gravity finally resulted in untilting Earth to its present state.

For Earth, being "knocked silly" conferred the kind of long-term stability that it needed. We won't always be so lucky. The evidence of the Moon's torque on Earth is plain for all to see simply by watching the ocean tides. But because Earth rotates on its axis once each day, and the Moon rotates around Earth only once every 28 days, the movement of the tides over Earth's surfaces dissipates energy. This dissipation causes Earth's own rotation to slow by 0.002 seconds per century, as the Moon gradually drifts away from us at a rate of 3.5 centimeters a year. As it does, it also loses its buffering influence over what would otherwise be our planet's unpredictable obliquity changes. As the Moon recedes from Earth orbit, our planet will succumb to gravitational interactions and perturbations from other planets in the Solar System over periods that range from 50,000 to a few million years. Some 1.5 billion years from now, the Moon will orbit Earth only 10 percent (or 40,000 kilometers) farther than it does today. Earth's rotation will become some 10 hours slower, making for a 34-hour day. And even though our orbit around the Sun will remain stable, Earth's rotational axis will become subject to large-scale (0 to 90 degree) obliquities triggered by the gravitational influences of Jupiter, Saturn, and Mars. By that point, discussion of such obliquities will be moot, but for now, these figures remind us just how tenuous habitable zones actually are.

1   Kasting, James F., planetary scientist, Penn State University. Interviewed on August 5, 1999, at Bioastronomy 99, Hawaii.

2   Canup, Robin, planetary scientist, Southwest Research Institute, Boulder, Colorado. Interviewed on March 21, 2001.

3   Laskar, Jacques, celestial mechanic, Paris Observatory. Interviewed on January 14, 2000, in Paris.

4   Interview with Darren M. Williams, planetary scientist, Penn State, Pennsylvania.

5   "One Theory Solves Two Ancient Climate Paradoxes." Penn State press release (December 14, 1999).

CHAPTER 9

In Pursuit of the Perfect Image

En route from Cannes to the Observatoire de Haute-Provence (OHP), I wasn't out looking for traditional Provençal tablecloths or homemade herbal soap. Unlike the tourists I was passing on this high-speed, winding three-hour "road rally" of a journey in October 1999, I was rushing to make a late afternoon meeting with a man who has been dreaming of directly imaging extra-solar "Earths" for more than a quarter century.

Antoine Labeyrie, who splits his time as a faculty member at the Collège de France in Paris and his corner office as Director of OHP, first "officially" broached the subject of imaging extra-solar planets in 1975. That's when he sent NASA a two-page document detailing his ideas about how that goal could be achieved with the agency's proposed Large Space Telescope. Labeyrie's idea was to use the NASA space telescope's initial 3-meter design for its main mirror along with a coronagraphic camera—an instrument that blocks out most of a star's light in order to see a faint planetary companion. This plan would have, in principle, made it possible for NASA's proposed space telescope to detect extra-solar planets after only a few hours of exposure. "In those days," says Labeyrie, "[NASA] didn't have much interest in such things and didn't understand my concept very well. I went to meetings every month or so and tried to convince them to make this Lyot coronagraphic camera, as it was called. But NASA wasn't interested."[1]

Instead, NASA asked the French physicist and optical engineer to join an instrument definition team for what would later be known as the Hubble Space Telescope (HST)—at $3 billion dollars, at the time the most expensive scientific instrument ever built. But within two years of joining the group in 1975, Labeyrie quit the team over a dispute about prelaunch testing of Hubble's primary mirror. NASA should have listened to Labeyrie, for his concerns would prove to be the project's major Achilles' heel. Labeyrie did not, however, give up hope that his coronagraphic camera would be able to fly on the space telescope. To that end he asked NASA to officially approach the then relatively young European Space Agency (ESA) about his idea. NASA agreed, and, essentially, put the coronagraphic idea in ESA's lap. ESA approved Labeyrie's plan and agreed to build it, with the caveat that, in exchange, NASA would give ESA observing time on the HST. By 1977, Labeyrie's coronagraphic instrument had been given the name High Resolution Camera. Although it could also be used for various other observations, its main purpose was to look for planetary disks around stars, brown dwarfs, and extra-solar planets.

But the HST's initial 1983 launch date was delayed to 1986, due to budgetary problems with Congress and a basic Hubble program reorganization. To make matters worse, the later

The Hubble Space Telescope in Earth orbit. (NASA and STScI.)

launch date fell victim to a terrible coincidence: the space shuttle Challenger blew up shortly after launch on January 28, 1986. Because the Hubble team had been counting on Challenger to eventually place it into its low Earth orbit, Hubble was put in an indefinite holding pattern following the tragedy. It was more than four years later when Hubble was finally launched by the space shuttle Discovery on April 25, 1990, and took up its 593-kilometer Earth orbit.[2]

At first, it seemed that the project's bad luck would continue. When the Hubble finally began its mission nearly a month later, a 30-second star exposure made it clear that the telescope's 2.4-meter main mirror had serious imaging problems. An investigation into the problem revealed that the mirror's Connecticut-based manufacturer had simply made it 2 microns (micrometers) too flat. "It was a question of matching the number one and two mirrors, and the matching was incorrect," says Labeyrie. "As a team member, I had asked [NASA] to do the [prelaunch] testing, and the answer they gave me was: 'It's too costly. You would have to test in a vacuum, and to test in a vacuum you have to have a huge vacuum tank.' I told them, you don't need a vacuum; just open a window in the lab where the telescope is and look at Polaris, the North star. That will suffice to tell you if the optics are good. That was one of the reasons I resigned from the team."

It would be three years after launch before the shuttle's first Hubble servicing mission successfully corrected the problems by mounting a repair package known as COSTAR (for Corrective Optics Space Telescope Axial Replacement). But the repair's new optics rendered the Labeyrie coronagraphic camera useless, and as a result, Labeyrie's dream of finding planets with Hubble went by the wayside.[3]

Labeyrie has not given up on his dream, however. He has simply turned in another direction: to designs of ground-based telescopic arrays that he can "tinker" with on his own turf in southern France. His vision includes implementing arrays of large optical telescopes—from 10 to 25 meters each—linked over distances of 10 to 20 kilometers. He believes by using sophisticated image recombination and processing, it will be possible to achieve the precision to directly image Earth-like planets around nearby stars.

THE DREAM OF INTERFEROMETRY

All telescopic imaging is, ultimately, limited by diffraction. A telescope's diffraction limit is the point at which a telescope can resolve any astronomical point in space. Even during a lunar eclipse, the Moon will not disappear all at once, but rather in slight oscillations caused by a diffraction pattern moving across the path of the observatory—which, again, is due to light's wavelike nature. For example, it was only six years ago that astronomers were first able to directly image the surface of a star other than our own Sun. That's when the HST made the first ultraviolet images of the surface of the supergiant M2 star Betelgeuse, some 131 parsecs away from Earth. It's an achievement comparable to being able to resolve (or differentiate) a car's headlights at a distance of 9,600 kilometers.

Labeyrie, however, is planning on doing something many times more difficult: he aims to directly resolve, or image, extra-solar planets from mere points of light, all in the glare of light from their parent star. The enormity of this undertaking can hardly be described. To image, for example, an individual 1,000-kilometer-diameter chunk of the Amazon rain forest at a distance of 10 parsecs would most likely require numerous individual telescopes linked over distances of at least 3,000 kilometers.

To achieve this feat, Labeyrie is now working on plans for a telescopic array called CAR-LINA (named for Carlina acanthifolia, an alpine flower native to the area). The design of this so-called hypertelescope involves a loose spherical mosaic of mirrors anchored to bedrock, and it is intended to span 200 to 1,000 meters. The only moving part of CARLINA would be overhead tracking devices that would hover over the telescope's mirrors in tethered, helium-filled balloons. These tracking devices would move to conform to the sidereal movements of the stars, and, according to Labeyrie, would be "moderately" sensitive to finding and imaging extra-solar planets. He intends to start experimenting with a small CARLINA prototype, which could be built for only a few thousand dollars.

SIDEREAL
Sidereal refers to the movement of any body, usually a star, in relation to a fixed point on the celestial sphere. In contrast to a 24-hour solar day, a sidereal day uses a fixed star as a reference point as Earth spins on its rotation axis. As seen from Earth, a sidereal day is the time it takes for a fixed star

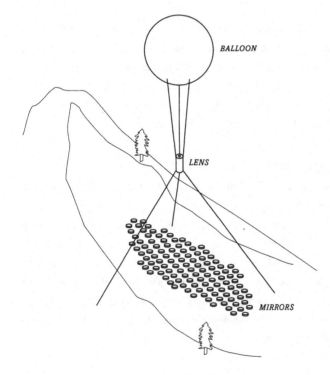

BALLOON

LENS

MIRRORS

Antoine Labeyrie's graphic conception of CARLINA, a revolutionary 200-meter hypertelescope whose spherical mirror elements would be solidly attached to bedrock. Labeyrie hopes the project will eventually see reality in southern France. (Based on a drawing by Antoine Labeyrie.)

to make a complete rotation around the celestial sphere: 23 hours, 56 minutes, and 4 seconds. Because Earth is moving along its yearly orbit around the Sun at the same time it is spinning on its axis, a solar day takes almost 4 minutes longer than a sidereal day. Armed with a given object's fixed celestial coordinates on the sky, right ascension (essentially the object's celestial longitude on the sky), and declination (or celestial latitude), astronomers can track the object for as long as it is above the horizon, assuming that the telescope is designed to track the object at sidereal rates. Such tracking is essential for viewing objects in narrow fields of view at high magnification. Otherwise, the target would very quickly move out of the telescope's field of view.

Labeyrie, with a team at OHP, is also studying plans for the Optical Very Large Array (OVLA) of 27 telescopes, which would span up to 1 kilometer. The array would move about to focus itself, and therefore concentrate its collected light at a common focus through a method of optical engineering known as interferometry (which is fully discussed in Chapter 11). As Labeyrie explains, the array's telescopes would reflect a narrow beam of starlight toward a central station to combine and form a single high-resolution image. The team has already constructed a 1.5-meter prototype OVLA element, which would be used in an initial version that would span a ring of 100 meters. An even smaller test version, dubbed "Micromegas," would involve 27 optically linked mirrors of only 20 centimeters each. OVLA's 1.5-meter elements, set atop hexapods, would be rolled into position along railway tracks. The telescopic elements could be ramped up in number from 27 to as many as 243 separate elements spread over distances as far as 20 kilometers. OVLA's success depends in part on finding a large and flat enough site to accommodate each of its 1.5-meter mirror elements. This larger, full-scale version would, in theory, be capable of directly imaging extra-solar Earth-like planets.

INTO THE FUTURE

The afternoon of my visit, Labeyrie and I talked until dusk. As night fell, a perceptible chill spread through his ground-floor office. Autumn had arrived with its longer nights, and the conviviality so evident at mealtimes around the observatory's dining tables in late spring had faded with the lavender. Michel Mayor's team was gone for the time being, and only three of the observatory's domes were scheduled for use that night. Sadly, while there, I learned that the observatory is in danger of closing. If not for Mayor's successful planet searches, the French government would likely have already closed most of OHP, opting instead to invest its time and efforts in the European Southern Observatory's new projects in Chile and elsewhere.

Labeyrie certainly knows the pitfalls of administering a large observatory, where decisions about funding can be made in the offices of those who have never spent time behind a telescope. But he remains undaunted in his efforts to realize his vision for the future, for he is driven by a keen interest in finding life "out there." Neither, as it turns out, has he given up on the Americans and their space telescopes. He has again approached NASA and ESA about funding a coronagraphic spectrograph for the Next Generation Space Telescope (NGST), now planned for launch by NASA in 2009. Labeyrie's proposed idea for a coronagraphic spectrograph could in theory be used to look for chlorophyll on extra-solar "Earths." Chlorophyll, a tell-tale signature of the life-sustaining process of photosynthesis here on our own planet, might have also evolved on other Earth-like planets in our galaxy. If so, it would be a convincing sign of extraterrestrial vegetation and possibly even higher forms of life.

"The coronagraphic spectrograph," says Labeyrie, "would be used for the visible and infrared and would give you an image of a faint dot showing the planet. Once you find the planet you can get a low-resolution spectrum from it, and if you are lucky, you will be able to see if the planet has green stuff." But for now, Labeyrie seems content to work on his ground-based projects. As I was leaving, he reminded me to be sure to look at one of his prototypical 1.5-meter OVLA telescopes, set in its futuristic-looking spherical hexapod. He needn't have worried, for later, while struggling to drive my way out of the observatory in the dark, I nearly slammed right into it.

1   Labeyrie, Antoine, astronomer and optical engineer, director Observatoire de Haute-Provence. Interviewed on October 16, 1999, at OHP, France.

2   North, John 1994. *The Fontana History of Astronomy & Cosmology*. London: Fontana Press: 594.

3   Fischer, Daniel and Hilmar Duerbeck 1998. *Hubble Revisited: New Images from the Discovery Machine*. New York: Copernicus Books: 26–28.

The Ancient Art of Astrometry
CHAPTER 10

The best hope of finding Earth-like planets around Sun-like stars lies not in Doppler spectroscopy, stellar occultation, water maser detection, or even direct imaging via telescopic ground arrays: it's in the ancient art of astrometry. Astrometry has long been a tedious and painstaking method of measuring a star's position, distance, and its proper motion (essentially its movement across our line of sight). In the second century B.C., without the benefit of binoculars or a telescope, the Greek astronomer Hipparchus spent years producing a catalog of over 1,000 star positions. Almost two millennia later, in 1718, Edmund Halley (whose name would later become inextricably linked with the comet) found that three stars—including Arcturus, a northern star that Hipparchus had recorded—had changed their positions on the sky by several fractions of a degree. Halley compared his findings against measurements compiled by Ptolemy, the Egyptian astronomer, who had assembled the work of Hipparchus and others in the *Almagest*, an astronomical compendium published two centuries after the birth of Christ. Halley's assumption that the stars must be moving across our line of sight marked the birth of modern astrometry.

Whereas Doppler spectroscopy measures a star's radial velocity shifts along our line of sight, movements which are then described in terms of meters or kilometers per second, astrometry marks out the circle that is our celestial sphere in terms of 360 arcdegrees. The sphere is further subdivided into 60 arcminutes per degree with 60 arcseconds per minute; thus, 1 arcsecond is 1/3,600 of 1 degree. For even more precise measurements, each arcsecond can be further delineated into hundredths, thousandths, or millionths of an arcsecond. At a distance of 10 parsecs, Earth would appear to be 0.1 arcseconds from the Sun.

PLANET HUNTING VIA ASTROMETRY

With enough precision, astronomers can also look for perturbations in a star's proper motion (its movement across our line of sight). Perturbations alone can signal the presence of unseen stellar or substellar companions, such as brown dwarfs or Jupiter-, Saturn-, or even Earth-like planets. Again, as mentioned in Chapter 3, as viewed from a distance of 10 parsecs from our Solar System, the gravitational orbital influence of Jupiter would cause our Sun to have a 500-microarcsecond astrometrical wobble. And again, Earth's own gravitational perturbation on the Sun (viewed from 10 parsecs away) would only cause the Sun to jitter by 0.3 microarcseconds. However, unlike Doppler spectroscopy, planet hunting via astrometry is inherently less sensitive to close-in planets such as 51 Pegasi b, and most sensitive to planets on multiyear, even decades-long orbits around their stars. So, although the rigors of astrometry have been used for centuries to facilitate navigation, its history as a tool for planet hunting has been fraught with uncertainty. But this uncertainty hasn't stopped planet hunters who use astrometry from making rather startling pronouncements through the years.

PROPER MOTION
The simplest way of detecting proper motion in a star is by recording its path as it travels across the celestial sphere. Like all movement in the Universe, an object's proper motion is relative to the observer. But unlike radial velocity, where objects move along the observer's line of sight, an object's proper motion is observed as it moves across (or perpendicular to) an observer's line of sight. Proper motions of stars are difficult to record without making repeated and often tedious comparisons with other stars in the telescope's field of view.

The late Kaj Strand, the first astronomer to publicly claim the "detection" of an extrasolar planet, did so when the rest of the world was more concerned about events here on Earth. In 1942, Strand based his claim on his observations of 61 Cygni, a binary star system 3 parsecs away in the Cygnus constellation, which he believed had an unseen companion 16 times the mass of Jupiter. Strand's claims were never substantiated. Astronomer Wulff Heintz, formerly of Pennsylvania's Sproul Observatory at Swarthmore College—where Strand had made his observations—told me he had studied Strand's 61 Cygni data, and that it had been subject to what Heintz termed "primitive" methods of analysis. (Nevertheless, Strand ended his career as scientific director of the U.S. Naval Observatory.)

Barnard's Star

Sproul Observatory had been a locus of astrometry ever since Peter van de Kamp, a Dutch astronomer, had arrived there in 1937 to conduct surveys of nearby stars. Van de Kamp became obsessed with Barnard's Star, the closest observable star from the Northern Hemisphere. Over several decades, he took thousands of photographic plates of the M5 star using Sproul's 61-centimeter refractor, a telescope that uses lenses instead of mirrors to magnify and produce images.[1]

The discovery of Barnard's Star was due solely to the diligence and passion of a self-taught astronomer. In December 1857, Edward Emerson Barnard was born into a dirt-poor family in Nashville, Tennessee. From a humble beginning as an assistant in a Nashville photographic studio, Barnard went on to become a world-class observational astronomer. In 1876, using only a $380, 5-inch (12.7-centimeter) refracting telescope, which he had bought from his own earnings, he discovered his first comet.

This success led to his being awarded a math scholarship at Nashville's Vanderbilt University. There, he made more comet discoveries on the university's 6-inch (15.25-centimeter) refractor. Then, in 1888, he took a day job inventorying property at California's Lick Observatory. It was here that he gained almost nightly access to Lick's 36-inch (91.5-centimeter) refractor telescope. By 1892, using the same Lick telescope, he achieved international fame for the discovery of Amalthea, the elusive fifth moon of Jupiter. Three years later, he took a job as a professor of practical astronomy at the University of Chicago's Yerkes Observatory, which gave him access to its new 40-inch (101.5-centimeter) refractor. During his career, Barnard took more than 4,000 photographic plates of our galaxy. Then, in May 1916, while examining a plate of a region in the Ophiuchus constellation, he saw that it contained a "new" star that he had never noticed before. Barnard examined an August 1894 photographic plate of the same region, taken during his tenure at Lick Observatory, to see if it also appeared there. This same "star" did indeed appear on this plate, but 4 arcminutes away from its 1916 position. A 1904 plate of the same region revealed yet another "new" star, almost halfway between the positions of the stars noted in 1894 and 1916. From this information, Barnard concluded that the images of the "new" star on all three plates were in fact pictures of the same star—although one with a very large proper motion. It was subsequently called Barnard's Star and still has the largest proper motion of any star known.[2]

At only 1.8 parsecs away, Barnard's Star is moving toward the Sun at 10.3 arcseconds per year, almost twice the proper motion of 61 Cygni, the so-called Flying Star. By 1963, van de Kamp had accumulated enough data on Barnard's Star to announce that it had a perturbation in its proper motion, which he claimed indicated that it had a planetary companion. He wrote that the suspected planet had an orbital period of roughly 25 years; furthermore, he claimed that this showed that the star harbored a 1.6-Mj planet, which he calculated would have a surface temperature of 60 degrees Kelvin. James Kaler, a UCLA graduate student at the time, was doing research for a thesis at a 1963 conference in Tucson, Arizona, when he hap-

Barnard's Star as pointed out with an arrow on two photographic plates taken in 1894 (above) and 1916 (below) by Edward E. Barnard. Barnard was captivated by the star's large proper motion, or movement across our line of sight, on the celestial sphere. (Photos courtesy Yerkes Observatory.)

Edward E. Barnard poses next to the Bruce telescope at the University of Chicago's Yerkes Observatory. (Photo courtesy Yerkes Observatory.)

Peter van de Kamp, a Swarthmore College astronomer, died convinced that there was a planet circling Barnard's Star. (Photo courtesy Friends Historical Library / Swarthmore College.)

pened upon van de Kamp's presentation on Barnard's Star. "He gave this riveting demonstration of the variable proper motion with the error bars, and it looked incredibly real," remembers Kaler, now a professor of astronomy at the University of Illinois at Urbana-Champaign. "As I recall, it was pretty quickly accepted that he had discovered a planet orbiting around another star. I don't remember what questions followed, but it was one of the few papers that I have remembered all my life."[3]

By 1969, van de Kamp had revised his initial claim. He had made what he termed an "alternative" analysis of the star's wobble. New data indicated that Barnard's Star had not just one companion but two. One had an orbital period of 26 years, and the other a 12-year period, with respective masses of 1.1 and 0.8 Mj. In 1975, van de Kamp "confirmed" the (now) 11.5-year planet. But the planet with the longer orbit remained "less well determined, at 22 years."

SHIFTING POSITIONS

The key to accurate astrometry begins with accurate alignment of the telescope's lenses. Notably, in 1949, during van de Kamp's decades-long survey of Barnard's Star and other stars, Sproul Observatory's 61-centimeter refractor telescope had undergone a major installation of a new cast-iron lens cell. To mitigate potential corruption of data in his survey, van de Kamp excluded data taken from 1938 to 1949. Nevertheless, other discrepancies were found in data from 1949 to 1956, and again in 1957, when the telescope's main lens had also been adjusted.

Despite van de Kamp's efforts to ensure the validity of his data, there was no denying that the adjustments had caused spurious data, which manifested itself in the form of slight shifts in star positions.

In 1971, George Gatewood, a young doctoral student at the University of Pittsburgh, wrote a thesis that effectively obliterated Barnard's Star's putative planetary system once and for all. Gatewood attributed van de Kamp's claims to bad data resulting from the telescope's adjustments. Heintz, who had analyzed almost 1,000 of van de Kamp's photographic plates of the star from the 1950s, also reports that many of van de Kamp's plates were woefully under-exposed: "The underexposure would tend to enhance a small displacement of the position of the star relative to comparison stars. And as was done at that time, if you select only three comparison stars, then you have no means to analyze what is an actual star displacement [caused by the perturbation of a planet], and what is a displacement caused by the photographic emulsion. The idea was to get as many plates as possible, even if they were of low quality."[4]

But van de Kamp and a few others remained intrigued by the possibility that Barnard's Star harbored a planet. In an effort to confirm van de Kamp's claim, Robert Harrington, an astronomer at the U.S. Naval Observatory, took more than 400 photographic plates of Barnard's Star using the observatory's 1.5-meter reflector telescope in Flagstaff, Arizona. Harrington was unable to detect a wobble in any of them. "There are many people who don't work themselves but still look for the mistakes made by others," van de Kamp told *Astronomy* magazine several years after his retirement. "I just let them talk. I have studied Barnard's Star for over 40 years, and in the last 10 years the interpretation has not changed."[5] Van de Kamp died in 1995, convinced that he had discovered at least one planet around Barnard's Star.

Four years after Gatewood had rained on van de Kamp's parade, the young astronomer made a U-turn, from his life as a planet obliterator to a new life dedicated to planet hunting. His redirection began with a phone call from John Billingham who, in 1975, was the head of Life Sciences at NASA Ames Research Center in Mountain View, California. Billingham was calling to invite Gatewood to give a talk about how to detect planets to a group called SETI.

"When he said 'Search for Extra-Terrestrial Intelligence,'" remembers Gatewood, "I said, "You know what? I'm pretty busy. [But] after Billingham read off the second Nobel laureate I recognized, I said that I'd be glad to come. About the third meeting, Billingham introduced me to David Black. One summer Black had a six-week study in which the different representatives in the field all sat together with the engineers and simply explored how we could do this. I think Black really nursed planet detection into its current state by pushing us to go back and look at it again."[6]

Measuring simple shifts in the proper motion of any given star requires a field of at least three stars—one target star and two acting as references. Then, on top of what Gatewood terms this "simple setting," he uses a field of 11 stars in order to model whatever else is going on in the field. In fact, he treats each star as if it were the target star, doing a new mathematical solution for each star every time it is observed. He then watches the movement of each star, which gives him an indication if there's anything unusual in his 36-arcminute field of view.

Gatewood, now director of the University of Pittsburgh's Allegheny Observatory, had been a high school dropout, but was drawn back to academia by astrometry's lure of statistics, instrumentation, and astronomy. He came to the University of Pittsburgh specifically to work with the Thaw telescope's lens, which has been used for high-accuracy astrometry since 1914. It is in that telescope's focal plane that today Gatewood uses a Ronchi ruling, or a piece of glass ruled with evenly-spaced lines spanning across the telescope's field of view. Each ruling line is about as wide as the star image. With each observation, the starlight passes into a microscope that is then focused into a fiber optic cable connected to a photomultiplier (an evacuated electronic tube that literally turns the stars' photons into electrical current and thus causes the release of individual electrons). That, in turn, enables Gatewood to know precisely how much light is coming from each star.

COUNTING STARLIGHT

Starlight is nothing more than electrogmagnetic waves that arrive in the form of photons, individual units or wavelets of light. It's unusual to find two stars that are exactly in phase (or whose wavelengths of light are in precise synchronicity). To account for this difference, Gatewood uses the photomultiplier to count each star's individual photons. From this number, he calculates the various stars' phase differences—which can be done to within a fraction of a Ronchi ruling. These fractions are simply figured into the total number of whole ruling lines already known to separate the individual stars, all of which allows for very precise astrometrical measurements. Explains Gatewood: "If it's a nearby star and it's moving through space pretty fast, then there is another motion that will be there. It's very much like someone throwing a baseball at you. At first you see the baseball coming and think it might hit you, but as it gets closer, you realize it's going to go over your head. You don't see much change in the acceleration angle, because when the ball is moving, its proper motion is small; but when it gets very close to you and goes directly over your head, the proper motion appears to be very fast as the ball whizzes by. That's an acceleration term called perspective acceleration, due to your position. We've got position, proper motion, parallax, and perspective acceleration, four terms that define where the star is."

Lalande 21185

Perhaps because they are seen as moving across the sky at faster rates, stars with large proper motions are observed more frequently and therefore attract all sorts of speculation. That may be one of the reasons that stars with large proper motions are frequently pegged as potentially harboring planets. The saga surrounding one such star, Lalande 21185, began some 200 years ago. The star had been cataloged by French astronomer Joseph Jérôme de Lalande in his catalog of some 47,000 stars, published in 1801 and reputedly based on observations made by his nephew. Today, the catalog is an astronomical footnote, but Lalande will forever be remembered for the 21,185st star on his list. It's a small M2 star 8.3 parsecs away from Earth in Ursa Major. Heading down into the galactic plane at a clip of 5 arcseconds per year, it automatically caught the attention of van de Kamp and Sarah Lee Lippincott, one of the Dutch astronomer's Swarthmore College protégés.

By 1944, van de Kamp had announced that Lalande 21185 had a variable proper motion. Sixteen years later, in 1960, Lippincott authored a paper claiming that Lalande 21185 was being orbited every eight years by a planet roughly ten times the mass of Jupiter. Gatewood investigated the Sproul Observatory's claims in the early 1970s. But after taking 143 exposures of the star at Allegheny Observatory, he found no evidence for a planet and noted that the variable proper motion of Lalande 21185, observed by both van de Kamp and Lippincott, was again due to systemic errors at Sproul Observatory. (Lippincott, who eventually succeeded van de Kamp, has since retired as director of Sproul Observatory. "If science is to progress," Lippincott responded when I asked her to comment on the observatory's history of premature announcements, "you have to put yourself out on a limb."[7]) But despite the systematic problems at Sproul, Gatewood noted in his 1974 paper that Lalande 21185 did have an unusual acceleration, an observation that he had failed to immediately pursue. "I can calculate what the acceleration should be if I know the distance and the radial velocity," says Gatewood, "and I was getting a wrong answer with Lalande 21185. That really puzzled me. With one exception, everybody who has ever studied this star has noted that the acceleration was wrong."

To try and resolve the dilemma, Gatewood pored over 50 years of photos, starting with collections taken in the mid-1930s, as well as eight years of observations with his Multichannel Astrometric Photometer (MAP). Finally, in 1996, he was ready to announce at a meeting of the American Astronomical Society that he suspected Lalande 21185 did indeed have a planetary companion. But instead of an 8-year orbit as outlined by Lippincott, Gatewood contended that the companion was on a 30-year orbit. Gatewood's data also showed a possible perturbation marking a shorter orbit, which he now believes could be due to a planet following a 6-year orbit around the star. Subsequently, a member of Marcy's team has reported that Doppler spectroscopy data taken from the Keck I telescope does not support Gatewood's data concerning the 6-year planet. Gatewood himself says that, while both planets have yet to be confirmed, he's not yet ready to write off his data regarding the 6-year motion. He's also willing to be patient regarding the possible 30-year variable motion.

"When we originally plotted out the Lalande 21185 data," remembers Gatewood, "we saw a very fast negative acceleration, a different acceleration from what you would normally have from the star's motion across the sky. We said, 'this is happening 30 years later, so it must be that the period is about 30 years.' Of course, you couldn't be exact about that because you don't know the exact true acceleration, and MAP measures it ten times more accurately than the plates did. But now we're seeing the acceleration beginning to reverse sines [or direction],

The 76-centimeter Thaw telescope at the University of Pittsburgh's Allegheny Observatory, equipped with George Gatewood's Multichannel Astrometric Photometer (MAP). Both the telescope and the photometer are mainstays of his planet-hunting efforts. (Copyright © H. K. Barnett.)

which is what is predicted from the photographic data. To me, that's the beginning of a confirmation, but I think it would be a little foolhardy to run forth and say, 'Yes we've confirmed it.' It's kind of job insurance: it might take me 30 years to get a whole orbit, so why not wait?"

THE FUTURE OF ASTROMETRY

In the time it takes to confirm a 30-year orbit, NASA or ESA could have already provided the answer from space. Surely, the future of astrometry lies in space. ESA's $700-million HIgh Precision PARallax COllecting Satellite (Hipparcos), named partly in homage to Hipparchus, was the first to prove that milliarcsecond astrometry was possible from space. Operating from a geostationary Earth orbit, Hipparcos continually scanned the entire celestial sphere using two separate focal planes that were sliced into 3,000 parallel slits. (These slits served the same purpose as Gatewood's Ronchi ruling.) In just four years, Hipparcos had measured the positions, proper motions, and parallaxes of 118,218 stars to an accuracy of 2 milliarcseconds. And it produced a secondary catalog of 1 million stars measured to within an accuracy of 20 to 30 milliarcseconds.

One of Hipparcos' primary goals was to measure the parallaxes of stars to the highest accuracies possible. The basis of calculating a stellar parallax is the fact that every six months, Earth moves roughly 300 million kilometers in its journey around the Sun. That's hardly a great distance in the universal scheme of things, but it is enough to give Earth-orbiting satellites and astronomers working at ground-based observatories enough of a triangulation, or "parallax

The payload of ESA's Hipparcos spacecraft in the pre-launch clean room. Hipparcos was subsequently launched in August 1989, and during four years of observations measured the positions, proper motions and parallaxes of more than a million stars. (European Space Agency.)

shift," to measure a star's tiny trigonometric shift against more distant objects in the background. By measuring this twice-yearly shift, astronomers can then calculate the distance to the target star.

Since the satellite's mission ended in August 1993, astronomers have continued studying data from its 17-volume catalog with intriguing results. For example, two researchers at the Observatoire de Paris report that they too confirmed that the short-period Jupiter-like planet circling HD 209458, the first extra-solar planet to be detected via stellar occultation, had also produced a similar dimming effect in the Hipparcos data. (However, at this writing, no other extra-solar planets have been confirmed using Hipparcos data.)

The next move is NASA's. The agency plans to launch the U.S. Naval Observatory's (USNO) $162-million Full-sky Astrometric Mapping Explorer (FAME) satellite in 2004. FAME will determine the positions, distances, motions, brightness, and colors of 40 million stars in our galactic neighborhood, to accuracies approaching 50 microarseconds. In its venerable 170-year-old star catalog, the Naval Observatory already has 500 million stars. But if FAME is successful, it will produce a catalog 20 times more accurate than Hipparcos—which will be nothing short of an astrometrical revolution.[8]

Claim to Fame

Using a solar-sail, rather than propulsion thrusters, FAME will reorient itself to scan the whole sky from a geosynchronous Earth orbit, looking in two directions at once, while rotating every 40 minutes. With a planned mission life of 2.5 years, which could be doubled after launch, FAME will scan the sky in a spiral, observing each star 2,000 times over the course of the initial mission time frame. And through its surveys, the satellite should be capable of detecting hundreds, if not thousands, of extra-solar Jupiter-like planets.

ESA, too, is making its next move. In 2012, it will launch a $500-million astrometric satellite as a follow-on to Hipparcos. Over the course of its five-year mission, Gaia, named for the ancient Greek goddess of Earth, will—according to an ESA mission summary—enable the disentanglement of "the six-dimensional space and velocity distribution of more than a billion stars in our galaxy and beyond." Gaia will be sent 1.5 million kilometers away from Earth, out of the way of eclipses, where it would continously scan the sky in two directions at a precision of 10 microarseconds—five times better than the best FAME can promise. In the process, it will discover tens of thousands of extra-solar planetary systems with accuracies equivalent to measuring the diameter of a human hair at a distance of 1,000 kilometers.

"The FAME design is basically a scaled-down Gaia concept," asserts Michael Perryman, who was ESA's Hipparcos project scientist and is now project scientist for Gaia.[9] ESA and the U.S. Naval Observatory have discussed sharing data and technology during both missions' planning phases via informal twice-yearly meetings. Adds Perryman, "I will not for one moment underestimate my American colleagues because they have a fantastic track record, and their system design is excellent. But however brilliant you are, it takes time to come up to speed on something like this. With the Hipparcos and Gaia system knowledge, I would say we have the experience in Europe and are very advanced in our thinking. With Gaia, we've got this great opportunity to push the understanding of our galaxy in a very fundamental way. I've been comparing it to the human genome project in understanding the human body, and I think with Gaia we will do something analogous for our galaxy."

Even so, FAME alone will offer astronomers 20 times the precision that Gatewood used to detect the variable motion of Lalande 21185. Pondering astrometry's progress, Gatewood recalled that when he started out, his thesis advisor told him that if he made a 20-percent improvement in precision over the course of his life's work, he would have his advisor's permission to say he'd really done a lot. "Since I got into astrometry," says Gatewood, "we've multiplied the precision by 10, and we're doing it again. Right now, radial velocity [Doppler spectroscopy] searches are finding planets and everybody wants to get onboard. But the next shift will be to astrometry."

1   Dick, Steven J. 1996. *The Biological Universe: The Twentieth-Century Extraterrestrial Life Debate and the Limits of Science*. New York: Cambridge University Press: 186.

2   "Barnard's Star." Department of Physics, University of Durham, U.K., http://www.dur.ac.uk/~dph0jrl/one_lab/pm_barn.html.

     Kelleher, Florence M. "Edward Emerson Barnard." http://astro.uchicago.edu/yerkes/virtualmuseum/Barnardfull.html (April 2, 1997).

3   Kaler, James B., astronomer, University of Illinois, Urbana-Champaign. Interviewed on January 14, 2000.

4   Heintz, Wulff, astronomer emeritus, Swarthmore College, Pennsylvania. Interviewed on March 23, 2001.

5   Schilling, Govert. "Peter van de Kamp and His 'Lovely Barnard's Star.'" *Astronomy* (December 1985): 26.

6   Gatewood, George D., astrometrist, Director of Allegheny Observatory, University of Pittsburgh. Interviewed on October 27, 1999.

7   Lippincott, Sarah Lee, retired director Sproul Observatory, Swarthmore College, Pennsylvania. Interviewed on May 10, 2000.

8   Seidelmann, Ken, Chairman of the FAME science team, U.S. Naval Observatory. Interviewed on May 25, 1999, at Dana Point, California.

9   Perryman, Michael, astronomer and ESA GAIA study scientist, ESTEC, the Netherlands. Interviewed on October 29, 1999.

Interferometric Beginnings

CHAPTER 11

In 1849, the same year that California pioneers were panning for one of the most "precious" metals a supernova can produce, Armand Fizeau, a little-known 30-year-old French physicist, was testing a cog-and-wheel-mirror contraption that was accurate enough to establish a "ballpark" estimate of the speed of light. Fizeau was clearly ahead of his time, for some 20 years after the California gold rush, he had already proposed a revolutionary way of making astronomical observations, one that is just now reaching full maturity. The young physicist proposed combining two beams of light to produce a barcode-like set of light and dark vertical fringes, which might enable astronomers to boost their telescopes' resolving power without having to increase their diameter.

FRINGE BENEFITS

Fizeau based his proposal on a classic experiment originally conducted at the turn of the nineteenth century by British physicist Thomas Young. Young had demonstrated the wave-like

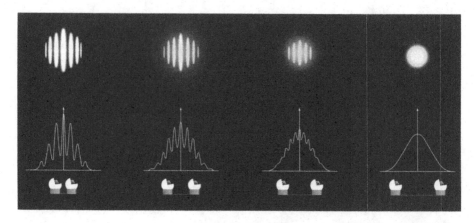

This graphic demonstrates the principle of interferometry, by which two separate optical telescopes combine light from a single star, resulting in interference fringes. As the distance (or baseline) between two telescopes is increased, the stellar fringes become less and less prominent. When the telescope's baseline reaches a certain critical point, the interferometric fringe patterns will disappear completely. As noted by Albert Michelson, the point at which the telescopic baseline results in the disappearance of these fringes is directly related to the angular diameter of the star. (European Southern Observatory.)

nature of light by showing that light, split into two beams and then recombined, would subsequently form interference fringes. The fringes are caused by the interference (or interaction) between wavelengths of light. When the crests of two wavelengths of light coincide exactly, they produce bright constructive interference fringes. When the troughs and crests of light-waves overlap exactly, the result is destructive, creating a nulling of the light as evidenced by dark fringes.

COLLIMATION AND NULLING
Nulling takes effect when two wavelengths arrive out of phase by 180 degrees with each other. In other words, the crests (wavelength peaks) and troughs (wavelength depressions) are electronically filtered and manipulated so that individual wavelengths of light cancel each other out, or combine destructively. When combining light interferometrically, contemporary instrumentation makes it possible to collimate the light from each separate telescope aperture. The photons are made to move in parallel to each other, so that when the streams of light are combined, their phase angles (literally, the angles of the various photon wavelengths) are synchronous. This allows astronomers to combine light more precisely, and by doing so, create interferometric fringes.

Today, Young's experiment is often repeated in many a high school physics class, but in his era, Fizeau saw that it also had possibilities for astronomy. The French physicist realized that by applying a mathematical formula in relation to the spacing of the wavelengths (or distance between the crests that span the light and dark fringes), it would be possible to estimate a star's angular diameter. Heretofore, astronomers had seen all stars as merely points of light; therefore, they believed the stars would remain impossible to resolve with any precision, much less have their diameters measured. French astronomer Edouard Stephan made the first attempt in 1874 by converting the Marseille 80-centimeter telescope into an interferometer. He did this simply by covering the telescope's aperture with a mask containing two adjustable small vertical slits, or openings. But he was still unable to resolve any stellar disks.

It was a Polish-American physicist at the University of Chicago who first succeeded in putting the method into practice. After visiting Fizeau in Paris in the early 1880s, Albert Michelson became inspired and returned to the U.S. determined to make the method work. In July 1890, Michelson attached a mask with two slits over a 12-inch (30.5-centimeter) telescope at California's Lick Observatory and, with a 10-centimeter variable baseline, measured the diameters of Jupiter's Galilean moons. Together with a colleague, Michelson spread the fringes apart by varying the baseline. They knew the wavelength of the light they were measuring and the distance between the light's interferometric fringe crests and could thus calculate the Galilean moons' diameter down to the arcsecond.

However, the method lay dormant for a few years while technology caught up—stellar diameters were simply too small for the embryonic technology to measure. Then, George Ellery Hale, an astronomer who had secured funding for three of the largest telescopes of the era, had the idea that Michelson's interferometry combined with the new 100-inch (2.5-meter) Hooker telescope (completed in 1917) might allow them to resolve the diameter of a newly theorized class of supergiant stars. Hale asked Michelson to mount an interferometer across the top of the giant telescope on Mount Wilson in the San Gabriel mountains of southern California.

The interior of the 100-inch (2.5-meter) Hooker telescope, which at the time of its completion in 1917 was the largest and most powerful telescope ever constructed. (Photo courtesy Scott W. Teare.)

In 1999, Albert Michelson's Stellar Interferometer, used on the 100-inch (2.5-meter) Hooker telescope, was removed from nearly 75 years in storage and put on display in CHARA's exhibit hall on Mount Wilson. (Photo courtesy Robert Cadman, CHARA.)

Michelson obliged, bolting on a 20-foot (6-meter) interferometer, consisting of two pairs of small, movable, half-silvered mirrors on a steel beam, which was strapped across the telescope's main aperture. Upon receiving starlight, the mirrors reflected it along equal path-lengths to the telescope's prime focus.

The Measure of a Supergiant

In December 1920, the new telescope, coupled with Michelson's interferometer, provided the first direct evidence that supergiant stars do exist. When measuring a stellar diameter, the fringes will eventually disappear as the baseline is increased (or the mirrors are moved apart). That moment of disappearance signals the relative measure of the star's angular diameter. Using Michelson's interferometer, Francis Pease, an American astronomer working at the observatory with Michelson, watched as the fringes from Betelgeuse disappeared when the mirrors were moved precisely 121 inches (3 meters) apart. By knowing the distance between the mirrors, along with the light's wavelength, the distance between the fringes, and the distance at which the fringes disappeared, Michelson was able to calculate the angular diameter of Betelgeuse. Lying 131 parsecs away from Earth in the constellation of Orion, Betelgeuse, an M2 supergiant star at least 5,000 times more luminous than the Sun, was found to measure in at 0.047 arcseconds. It became the first confirmed supergiant star.[1]

> *SUPERGIANTS*
>
> *Stars of at least 10 solar masses often evolve into supergiants after depleting much of their hydrogen and leaving the main sequence, which causes them to swell to great proportions. Betelgeuse and Rigel, both in Orion, are two of the most famous of the supergiants. Betelgeuse (also known by its scientific nomenclature as Alpha Orionis) is a red, variable star with a radius 600 times that of our Sun. Betelgeuse is already so large that if placed at the center of our Solar System, its atmosphere would*

extend beyond the orbit of Jupiter. Within several tens of millions of years, it is expected to explode into a supernova. In 1995, 75 years after Michelson and Pease made their historic measurements of Betelgeuse, the Hubble Space Telescope directly imaged this giant star's atmosphere, thus marking a new first: the first time that any telescope had been able to resolve the surface of a star other than the Sun.

Michelson was clearly ahead of his time. Computers were 30 years off, thus there was no effective way to track the stellar fringes as they danced across the sky. In another 75 years, these problems would largely be solved, but that would be too late for Michelson, whose attempts to measure smaller and smaller stellar diameters came to a halt in the mid-1930s.

INTERFEROMETRY ADVANCES

Interferometry continued to languish until after World War II, when the development of radio astronomy brought with it the capability to combine radio wavelengths. In the mid-1950s, Martin Ryle, a physicist at the famed Cavendish Laboratory in Cambridge, England, was among the first to show that it was possible to electronically store radio signals for a short period of time before combining them interferometrically.

Since then, radio interferometry has been done using arrays of antennas on the ground and even in the interferometric linking of ground arrays with antennas in space. However, interferometry at higher wavelengths is more challenging. Optical interferometric arrays of the sort that Labeyrie has advocated with his OVLA project have been much more problematic, simply because optical wavelengths are much shorter and, therefore, more difficult to combine to any precision.

Challenges notwithstanding, Labeyrie takes pride in being the first to bring optical interferometry into a new era. On the outskirts of Nice, at the Observatoire de la Côte d'Azur, he successfully linked two separate 25-centimeter telescopes in 1974 and was able to obtain interferometric fringes from the bright star Vega. Today, a larger version of his two-telescope interferometer, the GI2T (Grand Interféromètre à deux Télescopes) uses two 1.5-meter telescopes near its site at Calern in the southern French Alps.

Out of the Rafters, into the Future

In early May 1999, a construction crew working on a new multimillion-dollar interferometric array pulled Michelson's one-ton beam from the rafters of the 100-inch (2.5-meter) telescope dome where it had lain collecting dust for over 75 years. Three days after its resurrection, I arrived along with a group of astronomers from a NASA-sponsored conference on the future of space-based interferometry, to pay homage to Michelson, his interferometer, and the Hooker telescope, and to tour a small museum documenting the work of both Michelson and the observatory. At 1,740 meters, Mount Wilson arguably has the best "seeing" of any observatory in the continental U.S. The route to Mount Wilson winds around for miles on end before finally emerging into a mountaintop science reserve that has the look and feel of the old American West. It could just as easily be a nature reserve. Filled with towering pines and savage terrain, it's easy for visitors to forget they are little more than an hour's drive from the smog, traffic, and hubbub that is Los Angeles. Here the only hubbub comes from the occasional hoot of an owl, preying on mountain rats the size of small cats.

After marveling at the ancient interferometer, which at the time of my visit lay atop two sawhorses in a corner of the old observatory dome, we walked outside under a glorious blue sky, wandering into the interferometric paths of six 1-meter telescopes that make up Georgia State University's Center for High Angular Resolution Astronomy (CHARA) $13.5-million array. Constructing the CHARA array in such a wicked landscape was a major undertaking. When it came time to attach and bolt down the 5-meter high domes to their newly constructed bases, a helicopter was called in. Though hard to believe, Harold McAlister, an astronomer and director of the Georgia State group, assured me: "We installed six of these telescope structures one morning in less than an hour. The most amazing thing is that all the bolt holes matched perfectly."[2] The six automated telescopes are configurated along the lines of three radial 200-meter long arms that essentially lie at 120-degree angles. The result is an equivalent resolving power of a 400-meter telescope. The light from each telescope is sent to a central combining facility located in a building several hundred meters away from the array, where it arrives via vacuum pipes 12.7 centimeters in diameter. The light is then further compressed into pipes of only 2.5 centimeters in diameter.

Combining the light of two telescopes is a real feat, but combining the light from six approaches wizardry. The key to producing interferometric fringes is to make sure that the light

The construction of the six-telescope, $13.5 million CHARA interferometric array atop Mount Wilson was news-making in itself, as local television stations in nearby Los Angeles scrambled to get video footage of a helicopter positioning the telescopes' domes on their bases. Astronomers, however, are more interested in the array's optical interferometric resolving power—equivalent to that of one gigantic 400-meter telescope. (Photo courtesy M. Colleen Gino.)

from the telescopes arrives exactly in phase, so that the crests of the wavelengths match to within a few thousandths of a millimeter. To achieve this, the path-lengths themselves must be equal. Thus, the interferometer must account for path-length differences from the telescope to the light combination point, for the differences in Earth's atmosphere as viewed from each individual telescope, and for the fact that the star might be a tiny bit closer to one telescope than another.

At the CHARA array, once light has been fed into the three radial lines, it is sent back into a long one-story building, where it is manipulated by sets of steerable flat mirrors mounted atop small platforms that roll along tracks at high speed. Because, as mentioned previously, the light from each separate telescope arrives at slightly different times, these mirrors move along the top of parallel tracks in order to add length to each individual path of light. By adding path-length to each individual beam of light, the mirrors in fact time-delay each beam so that when it arrives at the central combination point, it arrives in synchronicity with the other individual beams of starlight. All this allows for the control of the light beam delay lines from each telescope down to accuracies of 10 nanometers over lengths spanning 240 meters (or over two and a half American-length football fields). From there, the light doglegs off into a beam combiner that will actually take the resulting interferometric fringes and recombine them into a "normal" image. Not all interferometers have the luxury of image production, as some simply rely on information that can be gleaned from the production of interferometric light and dark fringes, as when Michelson measured the angular diameter of Betelgeuse.

In late fall of 1999, CHARA successfully produced its first interferometric fringes. As this book goes to press, the array is well into its commissioning phase, which will last until the end of 2001, when it is expected to become fully operational for science. CHARA promises 200-microarcsecond resolution, enabling astronomers to see minute detail in objects (in principle, the equivalent of the accuracy that would allow astronomers to clearly resolve an American-length football field on Mars). While the CHARA array was designed primarily to do astero-seismology (the study of a star's global oscillations or variations), it will also be used to look for planets around binary stars. "We will look for submotions in a binary system [caused by] the presence of a planetary companion," says McAlister. "We will look for the reflex motion of Neptune-mass planets as they go around the binary star—I'm talking about a fraction of arc-second-narrow angle. The stars are so close together that you know they are physically related to one another. Most techniques blur the data and can't unravel the contribution of each individual star, but with an interferometer, we can. In [late] 2001, we will set up an observing program to search for planets in 50 binary systems within 100 parsecs [from Earth]."

Because CHARA will use such narrow angles, it won't have as many problems imaging through the roiling turbulent nature of Earth's atmosphere as do larger telescopes, such as the twin Keck 10-meter telescopes atop Mauna Kea in Hawaii. The Keck Observatory has undergone an interferometric upgrade that allows it to link its two massive telescopes. On March 12, 2001, it used the combined forces of both telescopes to obtain interferometric fringes of the

Harold McAlister, director of Georgia State University's Center for High Angular Resolution Astronomy in Atlanta, here on Mount Wilson in front of one of the CHARA array's six telescope domes. (Photo by Bruce Dorminey.)

faint, reddish M0 star HD 61294, some 250 parsecs away from Earth in the constellation Lynx. And eventually the Keck Observatory, with funding from NASA, would like to add four smaller 1.8-meter telescopes to the interferometer, which would enable astronomers to planet-hunt via astrometry—that is, by using the resulting interferometric fringes to look for the telltale wobble in a star's proper motion. However, the telescopes have yet to be constructed and do not yet have a fixed date for start of commissioning.

STAR WARS AND THE BIRTH OF ADAPTIVE OPTICS

Although the idea of adaptive optics as a method of mitigating the turbulence of Earth's atmosphere was first proposed in 1953 by American astronomer Horace Babcock, its development is linked to the so-called Star Wars research conducted as part of the U.S. Military's Strategic Defense Initiative (SDI) during the Reagan administration. The Pentagon needed to design lasers that could be sure of intercepting suborbital enemy missiles. However, unlike atmospheric disturbances that can interfere with the potency of a laser being fired at an enemy missile, astronomers simply need to make sure that the light they receive in their telescopes is as stable and free from atmospheric turbulence as possible. If astronomers can successfully model the atmospheric disturbances of Earth's atmosphere, then they will be better equipped to counter its deleterious effects. One way to mathematically model the effects of Earth's turbulence is to use a laser fiducial as a means of comparison. By using a laser to probe Earth's atmosphere first, astronomers can approximate their field of view's atmospheric turbulence. This allows them to determine the effects of such turbulence on stellar wave fronts (or groups of photons arriving in lines of synchronous waves) coming in from nearby stars.

As it turned out, the Pentagon's research to find a proper laser fiducial would uncover something even more useful: adaptive optics. Using adaptive optics, the atmosphere in a telescope's field of view is mitigated after it hits the telescope's focus. Although Babcock had first put forth this idea in 1953, its execution was hampered by lack of technology. That's because adaptive optics requires high-speed computer "gymnastics" in conjunction with mirrors that systematically match and remove turbulence in the wave fronts received from incoming starlight. Adaptive optics systems sometimes use real stars as reference (guide) stars, or an artificial laser fiducial to probe the atmosphere in the telescope's field of view and thus to measure the turbulence between the telescope and the celestial target. Once that is done, the adaptive optics system uses mirrors to counter the turbulence distorting the incoming celestial image, which keeps the image sharp, or pointlike. Amazingly, it works and is now in use in most of the world's latest technology telescopes.

When the Pentagon began experiencing targeting problems in its next-generation Patriot missiles, officials concluded that the best solution would be to use an infrared interferometer equipped with adaptive optics to determine where they were going wrong. Thus, in the late 1990s, the U.S. Department of Defense (DOD) approached Dave Westpfahl, an astrophysicist at the New Mexico Institute of Mining and Technology in Socorro, about using one of its sites in the Magdalena mountains that have been set aside by Congress for research. "The DOD wants to track advanced Patriot missiles over the White Sands Testing Range," says Westpfahl. "They would like a telescope overlooking the northern part of the range so that they can watch these intercepts and see the details of what happens."[3]

Pending congressional approval for the $40 million dollars required for the project, the DOD will build three 2.4-meter telescopes at the Magdalena Ridge Observatory, a site administered by Westpfahl and New Mexico Tech. Once construction is complete, the first interferometric fringes are scheduled to arrive in 2005. Westpfahl expects to use the new observatory to search for nearby planets, particularly planetary systems forming around what are known as T Tauri stars (stars less than 10 million years old), which are in the earliest stages of planet formation. The T Tauri stars' protostellar disks should be very easy for the group to see in the near and mid-infrared spectrums, because their radiation sometimes clears out circular, doughnut-shaped areas within which planets can begin to form. Westpfahl hopes to watch that happening.

Subtle Differentials

Meanwhile, a team of astronomers at the California Institute of Technology (Caltech) in Pasadena has scoped out time on the Keck Observatory to implement a whole new planet-hunting technique. They plan on using interferometry to determine shifts in a star's differential phase (or shifts in the angle of a given star's photon wave fronts—which could be gravitationally perturbed by planetary companions). If the team is successful in determining such shifts, they could then find planets.

Rachel Akeson, one of the Caltech astronomers, says that, beginning in 2002, she and her Caltech colleagues hope to use time on the newly interferometrically linked 10-meter Keck telescopes to look for known candidate extra-solar giant planets that orbit close to their stars, such as 51 Pegasi b, Tau Boötis b, and Upsilon Andromedae b. "A star is tens of millions times brighter than the planet," explains Akeson. "Because you have a small planet next to a bright star, you are going to get some deviation in the signal."[4]

All fringes have an amplitude and a phase component. The phase is the angle of the wave's amplitude (that is, the way the fringes slant from wavelength crest to wavelength crest). The phase contains information about where the source (or target object) lies on the sky. In a star-planet system, the planet is too faint to detect directly, so Akeson hopes to see the phase change (i.e., witness the perturbations in the star's wavelengths caused by the presence of a planet). She won't be able to measure the extra-solar giant planet's spectrum directly, as the British team tried to do when observing reflected spectra from the close-in planet around the star Tau Boötis, but she will be able to infer what such spectra look like by observing the differential interferometric phase shifts of the star and planet. By combining her own new data with what has already been gleaned about the close-in giant planets from previous Doppler spectroscopy studies, Akeson plans to use modeling to better determine the planet's temperature, chemical composition, and true mass.

The differential phase detection technique works best in the near infrared. But because differential phase-shift measurements are difficult to obtain, high-quality adaptive optics are a must. To that end, the team is working on modeling Earth's atmosphere as a whole, to better understand the turbulence that could affect their observations and thereby give them better ways of separating the noise from the signals. They are also studying new research on how atmospheric water vapor can harm their observations.

"Earth's atmosphere," says Roger Linfield, an infrared data analyst at Caltech, "will be one of the biggest error sources in confusing a differential shift in wavelengths of the stars and their planetary companions with the wavelength flux of Earth's atmosphere. What I can do is to quantify what the noise from the turbulence will be. Then I can understand what the noise is, so if [the astronomers] see something, they can determine if it's real or not."[5]

THE NULL FACTOR

Eighty years after Michelson had measured Betelgeuse's angular diameter, Phil Hinz, a graduate student from the University of Arizona, used a nulling interferometer on two 1.8-meter mirrors at the Multiple Mirror Telescope at Mount Hopkins, Arizona, to null the supergiant star. It was the first time nulling interferometry had been demonstrated on a telescope. He achieved a null factor of 25 to 1 in the mid-infrared, which meant that 96 percent of the star's incoming mid-infrared spectra were blocked out. The technique relies on the fact that each wave front is lined up exactly out of phase, or crest to trough, crest to trough.

Infrared nulling is one key to the success of the new $84-million Large Binocular Telescope (LBT), an American-Italian-German collaboration led by the University of Arizona and Italy's Osservatorio Astrofisico di Arcetri. Being built on 1.2 acres atop the 3,191-meter high Emerald Peak in the Pinaleno Mountains of southeastern Arizona, the LBT's eight years of construction have been hampered by access only via unpaved single-lane roads, its remote location within an endangered species refuge, and weather that typically halts construction from November to March. But once complete, the LBT should be able to use infrared wavelengths to directly image Jupiter-mass planets that lie only 0.3 AU from their parent stars. That's about the distance Mercury lies from our Sun. To directly image planets, the LBT must be able to null their companion stars by a factor of 10,000 in the mid-infrared, while maintaining its capability to image planets in extremely short orbits.

Though construction of the LBT has been a long arduous process, the first step in its construction went off without a hitch. The LBT's first 8.4-meter, 16-metric-ton honeycomb mirror was cast in 1997 at the University of Arizona's own Steward Observatory Mirror Lab using 21 tons of borosilicate Japanese-made glass. (Borosilicate has a durability not unlike glass wool or Pyrex and reaches a consistency similar to thick honey when melted, which made it possible to fit into its honeycomb mold of more than 1,600 hexagonal-shaped cores.) After polishing, it will finally see installation on the telescope in 2003. The second mirror will be installed a year later, and full scientific operations are scheduled to begin late in 2004. Mounted side by side, the two mirrors will have a total baseline of 22.8 meters, and the entire structure is designed to rotate

An artist's conception of the exterior structure of the Large Binocular Telescope (LBT) in the Pinaleno mountains of southeastern Arizona. The LBT will use the technique of interferometric nulling to achieve infrared direct imaging of gaseous giant planets circling nearby Sun-like stars. (LBT.)

for full sky coverage. NASA and ESA are counting on the Arizona telescope to smooth over some of the technological kinks that are sure to arise in at least three future space-based planet-hunting and imaging missions.

Astronomers are on the verge of using the principle of interferometry to achieve what Michelson and his colleagues at the Mount Wilson Observatory could only have imagined. Both NASA and ESA are in the midst of planning space-based missions that will use interferometry not merely as a means of measuring stars' angular diameters, but as a means of directly detecting a fraction of the extra-solar planets that must orbit these distant stars.

1   Leverington, David 1995. *A History of Astronomy: From 1890 to the Present.* London: Springer-Verlag: 132.

2   McAlister, Harold, astronomer and CHARA Director, Georgia State University at Atlanta. Interviewed on May 25, 1999, at Mount Wilson Observatory, California.

3   Westpfahl, Dave, astrophysicist, New Mexico Institute of Mining and Technology. Interviewed on May 26, 1999, at Dana Point, California.

4   Akeson, Rachel, astronomer at Caltech, Pasadena, California. Interviewed on May 24, 1999, at Dana Point, California.

5   Linfield, Roger, astronomer at Caltech, Pasadena, California. Interviewed on May 24, 1999, at Dana Point, California.

Working on the Fringe

CHAPTER 12

During a mid-morning break, the conference organizers herded some 100 astronomers into a semi-circle for the obligatory group photo. They clustered at one end of a five-star hotel pool overlooking a landscape reminiscent of an old *Dynasty* episode, replete with a hilltop view of immaculately manicured grounds and the harbor of Dana Point, a seaside oasis halfway between Los Angeles and San Diego. Dubbed "Working on the Fringe," this May 1999 conference was sponsored by NASA's Jet Propulsion Laboratory (JPL) in nearby Pasadena, and the "Fringe" in the title made reference to interferometric fringes, not what just a few years earlier had been the fringe nature of optical interferometry itself. With the scheduled 2009 launch of its estimated $870-million Space Interferometry Mission (SIM), NASA is taking Michelson's interferometric technology to the big time. And tony Dana Point made an ideal venue for JPL to promote NASA's ambitious space-based interferometry missions to the international astronomical community.

"As astronomers, we are not just dreaming about the nature of the Universe, we are doing something about it," says Geoffrey Marcy, who was also on hand at Dana Point. "The beauty of this meeting is that most of the talks are about future instruments, instruments that are going to cost a lot of money, but the results are going to be spectacular. Searching for planets, searching for life, understanding how our own Solar System fits in—I don't know of another field in which people are so excited about the future, 10, 20 years, even a century down the line. I take heart that the people are here because our discovery of planets has sparked interest in planetary systems in general."[1]

MISSIONS POSSIBLE

SIM is the first of NASA's two major space missions whose main emphasis is planet detection, to seek out planets within 10 parsecs of Earth that range in size from 2 to 10 Earth-masses. SIM will also be capable of detecting Jupiter-mass planets at a distance of 1,000 parsecs from Earth. And because the spacecraft will astrometrically measure a star's proper motion, it will be able to detect planets that normally escape the Doppler spectroscopy surveyors. It will also give the astronomical community the most accurate measurements of the orbits and masses of all the current planetary candidates, whether they have been detected via Doppler spectroscopy or astrometry. Finally, astronomers believe SIM will definitively determine whether stars such as Lalande 21185 and Barnard's Star actually harbor planets.

SIM's design consists of three simultaneously operated optical Michelson stellar interferometers that use 30-centimeter aperture mirrors (about the size of dinner plates) with a maximum baseline of 10 meters. While any one interferometer (or combination of two mirrors) observes stars that have been picked as science targets (or science stars), the other two interferometers will focus in on bright reference stars that will enable the science interferometer to

This graphic illustration shows the design of NASA's Space Interferometry Mission (SIM), now scheduled for launch in 2009. Among other science goals, SIM will survey Sun-like stars for the astrometric signature indicative of extra-solar planets. (NASA / JPL / Caltech.)

maintain its orientation in space. To effectively track the spacecraft's overall stability to about 10 nanometers, or roughly 10 times the diameter of a hydrogen atom, JPL is designing a laser-guided metrology (or high-precision measurement) system. This system will monitor the relative orientations and movement of the three interferometers as a whole to insure that Caltech analysts are able to correctly interpret and process SIM's data.

Seven boxes will be mounted on SIM's boom-like architecture, each of which will contain a mirror. Although the boxes won't move, any combination of two mirrors mounted on the boom can be used to acquire starlight from the same source, which in turn will enable interferometry. Once a field of stars is targeted, the light is fed into 4.5-centimeter-diameter delay-line mirrors inside the spacecraft, to ensure that the starlight reflected from each mirror will arrive in synchronicity. The starlight is then directed into a beam combiner, which adjusts the delay lines to create interferometric fringes from the star.

In this manner, the positions of the star's interferometric fringes are recorded on a detector. By observing the target star over time, interferometric fringes can provide astronomers with higher-precision proper motion data as they track the star's journey across the spacecraft's line of sight. After launch into an Earth-trailing solar orbit and a six-month calibration period, SIM will be ready for its five-year science mission, targeting a total of 10,000 stars. And because SIM will be controlled from the ground, astronomers will be able to change its target list to respond to new ground-based detections or to follow up on data it has already been sent back to Earth.

But all these plans remain just plans for now. "Until we get to the construction phase, we can be axed," says Stephen Unwin, SIM's deputy project scientist. "JPL is responsible for SIM's overall architecture, but no one has built an interferometer for space yet, so it's actually a learning experience for all of us."[2]

The Guiding Light

With its highest accuracies approaching 1 microarcsecond, SIM will set about measuring a science star's proper motion and thus look for the telltale astrometric wobble that may indicate the presence of an extra-solar planet.

Though SIM's interferometer will always have at least four reference stars within its field of view, SIM's controllers must make sure that the stars SIM is using as references are themselves stable and numerous enough to cover the entire celestial sphere. At least 12 reference stars will make up SIM's circular 15-degree overlapping astrometric tiles. They will be used to ensure that SIM can seamlessly acquire targets from any part of the sky.

ASTROMETRIC TILES
Astrometric tiles are simply a convenient way for SIM to break up the celestial sphere into manageable chunks, so that it can easily observe the stars it is interested in, while being assured of having known and reliable reference stars for comparison inside each tile. Each reference star within the 15-degree tiles are termed grid *stars by astronomers, because they make up an imposed astrometric grid of the celestial sphere. A good analogy is when an Earth-orbiting weather satellite*

takes a photo of a landmass on Earth, all it reveals is the geography that is before it. The satellite photo doesn't delineate between national or local political boundaries, so there is no frame of reference. Usually television weather channels impose political boundaries over the satellite photo, so that we have a reference for what the satellite photo is actually showing us. Astrometric tiles serve a similar purpose for SIM: they enable it to have a set and manageable frame of reference.

SIM will use these overlapping astrometric tiles to guide itself around the celestial sphere. Ideally, the spacecraft would have as many as 5,000 grid stars for its tiles, or again, about the number that can be seen from the Northern Hemisphere on a clear night. SIM, however, will have to get by with fewer (more like 3,000) against which to astrometrically track and compare its science targets. "You want SIM's grid made up of the most boring stars that you can find," says Ralph Gaume, an astronomer at the U.S. Naval Observatory. "You want stars that don't have any planets, don't vary, don't have sunspots [starspots]. There is no technique that we can attain now with that level of astrometric accuracy."[3] The best grid star candidates are K giants, because they are quite numerous, are found in all directions of the sky, and have radii 10 to 50 times larger than our Sun, which makes them intrinsically bright and easy to observe. Sabine Frink, an astronomer at the University of California at San Diego, has run data analyses on 30,000 K giants from the Hipparcos catalog, most of which were at distances of 350 to 550 parsecs. Frink and colleagues culled some 12,000 as preliminary SIM grid candidates.[4]

The Nulling Mission

SIM was initially designed to test optical nulling interferometry for the first time in space. But that portion of its mission has now been cut. However, such nulling will be important for future space missions and will subsequently be tested at ground-based facilities. Interferometric nulling involves "inserting" an additional 180-degree light-path delay into its interferometer in order to invert one lightwave into another. When light is combined interferometrically, it produces both white light and dark fringes, so that it looks like someone has superimposed bars over any given image. In nulling, one of the two incoming wave fronts of starlight has to be time-delayed. This ensures that when the incoming starlight arrives at the beam combiner, it arrives 180 degrees out of phase. The lightwaves' troughs are combined with its crests, and a dark fringe will cover the area of the image where the bright light of the star image would have normally appeared. This would thus permit astronomers to directly detect an extra-solar planet in the glare of its parent star.

Without such nulling, the $1.5-billion Terrestrial Planet Finder (TPF), NASA's ambitious follow-on mission to SIM, would never work. TPF must first block out a parent star's light in order to detect the Earth-mass planets that circle it. TPF's nulling problem has been likened to trying to locate a firefly in the near vicinity of a blazing searchlight. Its goal is to achieve a null of 1 million to 1, but a null factor of 100,000 to 1 will suffice to allow it to pick off planetary systems up to 15 parsecs away. If it can successfully do this, TPF will be able to determine the measurement of an extra-solar planet of Earth's size, temperature, reflectivity, and precise orbit.

TPF is scheduled for launch in 2012 into either an Earth-trailing orbit around our own planet or to the L2, one of five so-called Lagrangian points first calculated by French mathematician Joseph Louis de Lagrange (1736–1813) as equilibrium points in the Earth-Moon-Sun system. In the case of L2, both the Earth and the Moon would always remain between TPF and the Sun. During its planned five-year mission, TPF is to survey 150 stars for extra-solar Earth-like planets. Once the planets' orbital parameters and temperatures are known with certainty, TPF will begin surveying them spectroscopically for the presence of massive amounts of ozone (O_3), carbon dioxide (CO_2), water (H_2O), and methane (CH_4). (If these detections do nothing else, they would at the very least start a debate about whether the planet in question could also harbor life.) TPF's four 3.5-meter telescopes will operate entirely in the mid-infrared, as four independent free-flying space probes spread over distances between 75 meters and 1 kilometer. The four telescopes will relay their data to a fifth spacecraft, which will then collect and combine it for transmission back to Earth. To accomplish this feat, the five spacecraft must maintain their positions relative to one other to at least a fraction of a centimeter. Within two to three days of spectroscopic observations, TPF could detect carbon dioxide if it's there, indicating the planet has an atmosphere. Water would be detected next, and then ozone. Methane, however, lies just at the outer range of TPF's detection capabilities. To detect ozone or methane, TPF would need to observe an extra-solar planet for at least two solid weeks, equivalent to the time the Hubble Space Telescope spends on one of its deep field surveys. "By extraordinary good fortune, right in the mid-infrared, where we get the best contrast, is where O_3, CO_2, and H_2O each produce a big signature," says Roger Angel, an astronomer at the University of Arizona and a member of the TPF science working group. "Ozone is formed photochemically from oxygen and is not expected to be present in appreciable abundance in atmospheres from which oxygen is absent. If we saw a massive ozone absorption in the atmos-

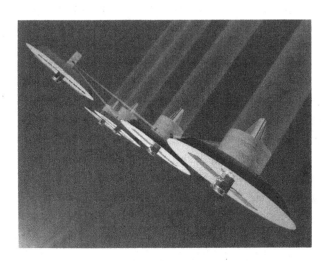

An artist's rendering of the reference design for NASA's Terrestrial Planet Finder (TPF). The spacecraft are shown to scale with a slightly shortened baseline. Now scheduled for launch in 2012, TPF will survey extra-solar Earth-like planets for the spectral signatures of large amounts of water, carbon dioxide, and ozone. (NASA / JPL / Caltech.)

phere of an exo-planet, then it would be better not to say that it's life, but to check it out. It would be a hell of an exciting thing to find."[5]

However, planetary scientist James Kasting of Penn State University argues that there are cases where oxygen and ozone could be detectable without the presence of life. Venus, for example, is a planet that clearly lies inside the inner edge of the habitable zone—too close to our Sun—and as a result has been the victim of a runaway greenhouse effect. Because Venus is closer to the Sun, it is also more susceptible to exposure from the Sun's ultraviolet radiation. That, in turn, causes photo-dissociation, a chemical process whereby the Sun's radiation has destroyed clouds of water vapor in Venus' upper atmosphere by "dissociating" the molecules' chemical bonds. This has resulted in massive amounts of hydrogen escaping into space, leaving large sources of abiotic oxygen, that is, oxygen that has been formed not by biological processes such as photosynthesis but as the result of Venus' temperature extremes. Methane, also a natural gas, is frequently produced as a biological by-product—as when it works its way through the belly of a cow. But because methane can be produced abiotically, by volcanic activity, for example, its signature could also be misinterpreted as that of life. To guard against interpreting such detections as a false positive for life, TPF would need to establish the orbital parameters and temperatures of the planets in question. This information would verify that the planet was indeed within the star's habitable zone.

As Kasting asserts, the spectroscopic detection of atmospheric gases, which, in combination, could indicate a planet capable of supporting or harboring life, should not be mistaken for hard-and-fast proof of life's existence. As a result, TPF data are certain to cause heated discussions at future astronomy conferences. "If you want proof," says Kasting, "you have to go there and pick it up."[6]

Future Financial Hurdles

TPF's first major hurdle will not arrive in the form of science, but in NASA's effectiveness in persuading the U.S. Congress to cough up $1.5 billion, possibly as much as $2 billion, to fund the project. NASA has always fared best in a race against the clock, as during the Cold War when it was challenged to meet President John F. Kennedy's timeline for landing a man on the Moon. But today, with SIM and TPF, NASA is contending only with its own visions for the future, so the agency's internal timeline will likely take a backseat to economics. Therefore, NASA will not ask for full TPF funding for several years yet. And, as Marcy notes, NASA has to convince Congress to look upon SIM and TPF as more than just "one-off" opportunities. He asserts that Congress should realize that NASA is on a quest, so that 50 years hence, we might pick up our newspaper and find a photo of a dot of light circling another star and then follow its orbital progress in the manner earlier generations followed Moon shots. "It would be glorious to directly image these planets," says Marcy, "but it's going to take the money and the effort."

Roger Angel, the University of Arizona astronomer, would first like to see NASA fund a scaled-down $80-million version of TPF in order to prove some of its technologies. "A mini-

TPF may not find a single 'earth,'" Angel remarks, "but one thing that we've learned from the Hubble Telescope is that we can go fuss with [the instrument] and tune it up. Directly detecting the light and heat from these planets is sufficiently difficult, [so] that some aspect of being able to learn and try again is going to be critical." However, unlike the Hubble Telescope, both SIM and TPF would be out of the Space Shuttle's orbital range if problems arose and servicing was needed.

While Angel may never get his mini-TPF, NASA is planning to launch a two-spacecraft formation-flying interferometry test mission in early 2005. The two spacecraft will prove that free-flying spacecraft can do precision optical interferometry in the visible spectrum. From a solar Earth-trailing orbit, the StarLight mission's two spacecraft will observe some 100 stars during its four- to five-month interferometry phase, using maximum baselines of over 1 kilometer. "Putting a man on the Moon was a trivial task compared to TPF," says Neville Woolf, an astronomer at the University of Arizona and the TPF preprogram mission scientist. "We haven't had a single interferometer in space yet, and we are suggesting that we go one-two-three, and TPF is the third interferometer in space. We don't learn that fast. We may be incredibly lucky, but I suspect that the program will get spread out. If you actually take history as the best guess, you can say that TPF is not 12, but 30 or 40 years away. In the realm of studying planets and life, we haven't gotten the concept. The goals are good, but hopelessly optimistic in the time scales for [accomplishing them]. The public doesn't accept failures, but failures are absolutely essential. But as we've got it, it's a set up designed for failure."[7]

DUST AND MIRRORS

If TPF does fly, ideally, it would "set up camp" beyond the orbit of Jupiter, where space is three times darker than at either of its currently planned on-station positions. But in order to send TPF out beyond the orbit of Jupiter within a reasonable period of time, NASA would have to appease protestors here on Earth who oppose nuclear-powered spacecraft because of the perceived environmental risks in the event of a malfunction during launch. However, from its vantage point near Jupiter, TPF could make the same observations even more effectively with spacecraft that carry telescopes of only 1 meter each, as compared with NASA's current design of 3.5-meter telescopes. NASA's Marshall Space Flight Center in Huntsville, Alabama, has set up an optics manufacturing lab which is working on revolutionary thin mirrors that would decrease the weight of a Hubble-type telescope by a factor of 100. These new optics will likely be ready and usable for TPF by the time it is ready to launch.

But problems remain. "Zodiacal dust is the primary nemesis of direct imaging of exoplanets," says Marcy. "That's a major bogeyman. Once we've overcome the star nulling, then we've got zodiacal dust, which is like looking through a windshield in fog. It has glare everywhere." Zodiacal dust in our Solar System is left over from its formation disk, but it also remains in other planetary systems and is most visible in the infrared. In some cases, exo-zodiacal dust in another system could be as much as 100 times brighter than the planets TPF would be trying to study. The Large Binocular Telescope in Arizona will survey as many stars as possible in advance of TPF's launch to effectively determine how much exo-zodiacal dust is out there, as well as act as a good proving ground for TPF's capability to study other planetary systems. NASA has also set up an official joint task force with ESA to look into combining TPF with ESA's nascent planet sensing program, the Infrared Space Interferometer/Darwin mission. (ESA is investing roughly $15 million dollars to study Darwin's prospects, and at some point, TPF and Darwin will join into one team.) In contrast to TPF's four telescopes, Darwin's current plans call for six to promote a more stable system. ESA's flotilla would include two additional spacecraft for data transmission and beam combining and interferometry. According to Malcolm Fridlund, the IRSI/Darwin study scientist, ESA's biggest problem remains in effective nulling of the starlight.[8]

Antoine Labeyrie, meanwhile, hopes that when the TPF and Darwin study teams decide to finally merge, they will abandon their current designs for TPF and Darwin and instead take *him* up on his own version of a planet finder, the Exo-Earth Discoverer (EED). Labeyrie's design calls for extra-solar spectroscopic planetary analysis, but on a much grander scale than anything being currently being considered by NASA or ESA. His planet finder, a 100-meter hyper telescope in space, would operate using 36 free-flying telescopes, each with mirror diameters of 60 centimeters. Instead of using interferometric nulling, Labeyrie proposes using a coronagraph (an artificial occulting disk) to remove starlight from the hyper-telescope's combined image. He estimates that in the mid-infrared, his instrument would offer planet–star contrast that is significantly better than what NASA and ESA proposed in their original mission concepts. Labeyrie believes that, at a distance of 3 parsecs, the EED could actually null the Sun and produce an image of the Earth and Moon in the hue of a yellow light. Labeyrie has had some success promoting his planet finder: NASA has included Labeyrie's EED design in a nine-month TPF industrial design study whose objective is to select some of the best ideas from several different concepts.

For its part, ESA has yet to incorporate Labeyrie's ideas into its proposed Darwin mission, although Fridlund says that it may happen. In any case, the laborious process of merging the two agencies' design teams and programs is likely not to come about for several more years. The reality is that much work remains to be done before the advent of space-based interferometry of any stripe. Marcy is the first to caution that no Earth-bound astronomers need to adjust their observations based on the advent of space-based observatories. He recalls a late-night conversation with his advisor, George Herbig, who, Marcy says, taught him everything he knows about observing at the telescope: "Most of his career was spent taking spectra of stars using photographs—no CCDs, no silicon detectors, just these grainy fuzzy little spectra. One night in 1979, we were at the telescope, and I asked him, 'How do you feel about your career? You have been working since the late 1940s taking these old spectra, which are now considered terrible. How do you feel that you spent hundreds of nights at the telescope taking data that is now clearly obsolete?' He said, 'That's what science is about. You spend your career doing something, working really hard, making a little progress, and if you're lucky, the results of your experiment point in the direction where we should go with the next technology. But your data will be rendered obsolete no matter how good an astronomer you are, and that's what we all have to live with.' I've taken Herbig's words to heart. We are working until midnight every single night. We work weekends, holidays, every day, because we are finding more planets than anybody. Why should we quit? I'm here at this [Dana Point conference] to help SIM and TPF be the best they can be, and if we are going to be rendered obsolete by these missions, I'll be as happy as anybody. I don't mind if someday, when I'm an old fogey, someone says, 'Ah, there's Marcy: he used to do detection of Jupiter-mass planets. Isn't that mundane?'"

1   Marcy, Geoffrey, astronomer, University of California at Berkeley. Interviewed on May 25, 1999, at Dana Point, California, and on August 6, 1999, at Hapuna Beach, Hawaii. Follow-ups took place on September 8, 2000, May 10–12, 2001, and June 3, 2001.

2   Unwin, Stephen, physicist and astronomer Jet Propulsion Laboratory, Pasadena, California. Interviewed on May 24, 1999, at Dana Point, California.

3   Gaume, Ralph, astronomer, U.S. Naval Observatory, Washington, D.C. Interviewed on May 25, 1999, at Dana Point, California.

4   Frink, Sabine, astronomer, University of California at San Diego. Private communication with the author on November 1, 1999.

5   Angel, Roger, astronomer, University of Arizona at Tucson. Interviewed on June 1, 1999.

6   Kasting, James F., planetary scientist, Penn State University. Interviewed on August 5, 1999, at Bioastronomy 99, Hawaii.

7   Woolf, Neville J., astronomer, University of Arizona, Tucson. Interviewed on August 5, 1999, at Bioastronomy 99, Hawaii.

8   Fridlund, Malcolm, IRSI/Darwin study scientist. Interviewed on August 4, 1999, at Bioastronomy 99, Hawaii.

Power Telescopes 13

Antofagasta, Chile's largest northern city, bills itself as the copper-mining capital of the world. Only 225 kilometers from the world's largest open-pit copper mine, the city has the look and feel of an old-world, hardscrabble port town that's seen its share of broken promises. The advent of fiber optic cable as a replacement for copper telephone wire could have dampened the town's spirits. But whatever its current prospects, this outpost of the northern Atacama desert is where it feels like the end of one world meets the beginning of another. For Antofagasta is the prinicpal gateway to the world's largest optical telescope—ESO's $500-million Very Large Telescope (VLT), a pantheon of fiber optic technology.

On a clear weekday July 1999 morning in Chilean winter, I found myself in an ESO minivan loaded with VLT personnel on a 120-kilometer trek from Antofagasta to the desert mountain of Cerro Paranal. From its lone vista in the midst of the driest desert on Earth (with only 5 percent humidity), this 2,600-meter mountain boasts the best astronomical seeing of any currently known site in the Southern Hemisphere. After tracing the coastal highway a few kilometers south, we turned into the ash-brown foothills of the mountain range that mark the entrance to the Atacama plateau. The only sign of civilization is a Bolivian rail line used to haul copper on a two-day journey between La Paz and Antofagasta. A few minutes later, we veered south onto the Pan-American highway for about 50 kilometers, which finally led us to a sharp turn onto 70 kilometers of the most rutted, rugged, awful dirt road imaginable. About halfway through this desolate stretch the vibrations were so extreme that my whole head throbbed in the manner usually caused only by extended encounters with a dental drill. That sensation was soon exacerbated by the dust sifting into the van ventilation's system and, inevitably, into my mouth.

At first glance, this stony desert seems absolutely immutable, devoid of anything resembling life. But, in truth, the region is primarily a result of 250 million years of raging tectonics. The neighboring Andes mountain chain was formed by the clash of two geological plates as they rode the molten surf underneath Earth's crust. The region's tawny hue stems from an eons-long oxidation process involving silicate rocks that exploded onto the landscape in great volcanic bursts. Today, this 700-mile-long desert remains seismically active.

A TECHNOLOGICAL OASIS

A lone sign stands at the entrance to an ESO-built two-lane blacktop heralding the last 12 kilometers leading up to Cerro Paranal mountain and the VLT. Our first stop was at the VLT base camp, a series of temporary prefab office buildings hastily constructed with corrugated aluminum, clearly indicating that the observatory was still a work in progress. Set against such a bleak landscape, the place had the uncanny look of a lunar base. On this particular afternoon, it played host to around 300 people, an eclectic international mix of astronomers, engineers, cooks, carpenters, bulldozer operators, ditch-diggers and dishwashers. Just to support the daily efforts of this diverse group, diesel generators pump out 300 kilowatts an hour, and three tanker truckloads of water are used per day.

Upon completion, the VLT site will include a $9-million, 7,600-square-meter complex built into the lower side of the observatory mountain, and constructed of tawny-colored bricks to match the color of the surroundings. This complex will contain a 140-room hotel, conference rooms, an administration center, and an indoor garden oasis and swimming pool that uses recycled water and electricity from the VLT's own water and electric plant.

Power Structure

After lunch at the base camp's *cantina*, I joined Jason Spyromilio, an ESO astronomer and head of VLT commissioning. We sped up the mountain in one of ESO's pickup trucks and pulled into a parking spot outside the VLT's control building, which lies several meters below the top of the mountain. This two-story ultramodern structure looks more like a corporate office building than the control center of one of the world's most expensive and modern instruments of science. Taxpayers in ESO's member nations might look upon the idea of funding such a complicated telescope with as much a sense of alarm as awe. Spyromilio acknowledges such concerns, but dismisses them as a small price to pay for the VLT's role in opening new vistas on the Universe. "Looking at the edge of the Universe isn't like mining Teflon as a [technological] spin-off of a manned space program," says Spyromilio. "The optical technology may have some commercial spin-offs, but not many. And looking for planets? Aside from curiosity, for most people you might as well be stalking monarch butterflies."[1]

The VLT project began in the early 1970s as an outgrowth of ESO's interest in building an instrument that would have the collecting power of at least a 16-meter telescope. But the technology to make a 16-meter mirror was unavailable at the time. So when Antoine Labeyrie suggested in 1974 that light from very large telescopes could be combined for interferometry, ESO took his idea of telescope combination and ran with it. Ten years later, ESO approached Marco Quattri, an Italian nuclear engineer—who, ironically, admits to having no interest in astronomy—and offered him the job of overseeing the VLT's structural mechanics. ESO's basic idea was to build four 8.2-meter telescopes into a trapezoidal configuration at an ideal site in Chile. Cerro Paranal mountain was chosen, because it gets up to 350 clear nights a year and there is so little water vapor in the local atmosphere that it is perfectly suited for infrared astronomy. Now the VLT's project manager for structural mechanics, Quattri says that he started from a blank sheet of paper in 1985, and by 1990 had the final plan.[2]

After seven years of site testing atop the mountain, ESO began blasting, and by 1991, had removed 350,000 cubic meters of rock and soil from the top of Cerro Paranal. The objective was to make a smooth, 20,000-square-meter site for the four telescopes, the auxiliary telescopes, and the underground interferometric labs. In honor of their host country, ESO gave each of the four 8.2-meter Unit Telescopes (UTs) names in the language of the Mapuche, an indigenous people from southern Chile: Antu (the Sun), for UT1; Kueyen (the Moon), for UT2; Melipal (the Southern Cross), for UT3; and Yepun (Venus), for UT4. Antu saw first light in late May 1998, and the president of Chile officially inaugurated the VLT in April of 1999. By 2001, all four unit telescopes were up and running. Three smaller 1.8-meter telescopes are due to be installed on the mountain by 2003. A year later, the VLT will also have a 2.5-meter survey telescope for target finding.

Each 8.2-meter telescope was designed to be installed in its own separate thermally controlled building, whose domes rotate as each telescope unit tracks its targets across the sky. The spherical design of the telescope domes was tested in a wind tunnel, to ensure that they could withstand gales of up to 100 kilometers per hour (even though the domes are closed when winds hit 50 kilometers per hour). Each telescope unit enclosure is air-conditioned both to maintain a uniform cooling of the telescope's components and to sustain a positive (or higher) air pressure inside the building than outside the telescope's enclosure, ensuring only a minimum of dust contamination.

Sunset over the four 8.2-meter Unit Telescopes at the European Southern Observatory's Very Large Telescope (VLT) in Chile's Atacama Desert, with the Pacific Ocean visible on the horizon. In this new decade, the VLT is expected to play a major role in all aspects of planet-hunting. (European Southern Observatory.)

Mirrors on the Universe

The telescopes' main 8.2-meter Zerodur glass mirror units were produced at Schott Glaswerke in Mainz, Germany, and were then transported by barge to France for polishing by a high-tech company south of Paris. After polishing, the mirrors were put into a specially designed shock-proof compartment and loaded aboard a ship bound for Antofagasta via the Panama Canal. From Antofagasta, each 23-ton mirror had to make the bone-rattling journey out to Paranal. The delivery of a $10-million irreplaceable piece of glass could never be routine, but ESO had it down to a science. Upon arrival in Antofagasta, the mirrors were carefully offloaded onto a 16-axle truck (each axle with an independent suspension system). The speed limit was 3 kilometers per hour, a slow amble. Three road graders went ahead of the truck to smooth out the bumps, and road inspectors walked behind each of the truck's wheels to make sure nothing unexpected could shake things up. The 70-kilometer "mirror run" from the Pan-American highway took some 24 hours, and when night fell, the delivery crew simply stopped in their tracks and waited for daylight. Once at Cerro Paranal, the mirrors were coated with aluminum in a vacuum chamber, a dust-free, particle-free "clean" room that is part of a permanent mainte-nance facility a few meters from the base camp. Finally, the mirrors were transported up the mountain, and with the help of a 300-ton crane, lifted into their ultimate resting places. To min-imize damage from earthquakes, each 8.2-meter telescope moves on its own independent sus-pension system that is totally separate from the dome's supporting structure. That way, the telescopes can stand earthquakes up to 7.8 on the Richter scale, almost the magnitude of the earthquake that devastated San Francisco in 1906. But ESO is not taking any chances: anything over 6.0 on the Richter scale will trigger the release of air bags under each of the main mirrors.

Transporting the first 8.2-meter Zerodur mirror to Cerro Paranal mountain, December 9, 1997. This photo was taken on the last stretch of road just before the base camp entrance. The VLT enclosures at the top of Cerro Paranal can be seen in the background. (European Southern Observatory.)

BIG-TIME ASTRONOMY

During a late afternoon tour inside the domes, it was hard not to notice that the 8.2-meter telescope units seemed more akin to electronics-filled industrial warehouses than isolated lone apertures pointing into the heavens. In addition to adaptive optics and all manner of instruments, each unit is also equipped with active optics, a system first pioneered by ESO at its La Silla observatory in the 1980s. Active optics involves tiny computer-controlled actuators, or supports, on the back of the mirror itself. These supports reshape the mirror to counter gravitational distortions as it is refocused to track its target across the sky. (Adaptive optics, by contrast, manipulate the target's incoming photons—to match and mitigate their inherent imperfections in image quality—after the incoming photons have already been reflected by the telescope's main, or primary, mirror. This is usually done with a computer-controlled deformable mirror that is located behind the main focal point of the telescope). Active optics manipulate the primary mirror while it is in the process of receiving images, to counter gravity, and to enable the mirror to maintain its alignment and shape. On the 8.2-meter telescopes' primary mirror, this generally means that about once each minute, the actuators located on the back of the main mirror exert a correcting force on the back of the main mirror itself. This allows the VLT's 8.2-meter unit telescopes to acquire a source anywhere on the sky within three minutes and then observe it with a resolution 4 billion times better than the naked eye. "If you're sitting here in Paranal, you can actually resolve a pair of headlights on a car a thousand miles away in Santiago," Spyromilio says, sliding into a chair at one of four horseshoe-shaped consoles inside the VLT's main control room.

Spyromilio explains that because the telescope is so powerful, the background light from the sky in the telescope's field of view grossly exceeds the amount of light of almost any object the VLT observes. He compares it to looking at a candle held in front of the Sun through the screen of Earth's atmosphere. Because this screen is moving and the telescope automatically

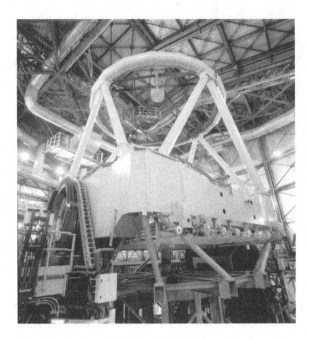

Inside the closed dome of the Yepun (UT4), the last of the VLT's four 8.2-meter Unit Telescopes. The base of the telescope in the foreground holds the main 8.2-meter mirror. (European Southern Observatory.)

corrects for this effect with at least three simultaneous movements, Spyromilio says that it's almost as if the telescope itself is experiencing some sort of debilitating seizure. In other words, the VLT and all its requisite technology would quickly overwhelm any astronomer who wasn't accustomed to working with state-of-the-art large aperture telescopes. As Spyromilio points out: "This is not, 'Let's go out into the backyard, stick our eye behind a telescope, and off we go.' You're not just going to let someone in here to look at a star and say: 'Oh, I saw a star. It was bright and it was blue.' And then, maybe if they were really conscientious, write it down. This is big-time astronomy. People expect a serious effort."

A NIGHT AT THE TELESCOPE

Walter Jaffe, an astronomer at Leiden Observatory in the Netherlands, spent his nights at the VLT looking for the signatures of molecules in the distant intergalactic medium. "I'm trying to detect what's in the intergalactic medium," Jaffe says, clutching a magnetic tape with data from two nights of observing on Antu, one of the 8.2-meter telescopes. "I have been at quite a few observatories, and this is the nicest I've ever seen it—the seeing and the darkness; it's really dark. But it's very complicated, and there are all kinds of things that you have to know about it. It's very easy here to make mistakes, and they don't know all its quirks yet."[3]

Jaffe reviewed his previous night, which he had spent looking for intergalactic gas emission lines in Abell 2597, a cluster of at least 50 galaxies some 500 million parsecs away from Earth. (George Abell, a UCLA astronomer who took most of the pictures in the first version of the Palomar Observatory Sky Survey, did a survey of the entire northern sky. Most of the galactic clusters that are in Jaffe's current target list come from Abell's 1958 catalog.) Emission lines are the hardest of such spectra to spot, as they are created by excited atoms of gas that release photons into the surrounding void. Though difficult to detect, their characterization in the intergalactic medium gives theorists an idea of the Universe's basic chemical abundances. In his own data, Jaffe found spectra from oxygen, nitrogen, sulfur, and hydrogen, but no iron. "The current theory of the Big Bang is that when the Universe was very, very hot, there were no atoms as we know them, just protons, electrons, and neutrons," says Jaffe. "When it cooled, the collisions between the three could form hydrogen, helium, deuterium, and lithium. But when you try to get beyond lithium to the heavier elements, they are relatively unstable and tend to fall apart fairly quickly. The first-generation stars were probably only made of hydrogen, helium, and lithium. Eventually, as the cores of those stars burned their hydrogen in thermonuclear reactions, eventually the heavier elements like nitrogen, carbon, and oxygen were formed, and these keep burning until you get to iron. I think the iron to form the planets is available around the [whole] galaxy in the same way it's available around our Sun. The fact that we haven't found smaller planets is just that they are very hard to find, not that they are not there."

"Big time" for the VLT and its interferometer, in part, means the capability to planet-hunt with all the currently known methods:

- using the 8.2-meter units for traditional Doppler spectroscopy;

- disentangling the spectral signature of a planet in the Doppler reflex motion of its parent star, as the British attempted with Tau Boötis b;

- astrometric planet detection using a single telescope equipped with a Ronchi ruling, as in the method of George Gatewood;

- astrometric detection via traditional interferometry and differential phase-shift detection between contrast changes in interferometric fringes (as proposed by Caltech for use on the Keck Observatory in Hawaii).

Finally, the new facility will attempt direct imaging of extra-solar planets in the near-infrared through the use of a coronagraphic mask in the focal plane of an infrared camera. The VLT will also have a spectrometer with the sensitivity to allow for Doppler spectroscopy on very faint stars, such as the M star Proxima Centauri, Earth's nearest stellar neighbor.

THE IMPORTANCE OF VLT INTERFEROMETRY

The VLT's real planet-hunting cachet lies in its interferometric capabilities. With four 140-meter subterranean interferometric delay lines, the VLT is capable of interferometrically linking all four large 8.2-meter apertures at once. The most likely configuration, though, will be one 8.2-meter telescope linked to at least one, perhaps two, of the smaller 1.8-meter auxiliary telescopes now under construction in Belgium. In theory, the VLT Interferometer (VLTI) will have a resolution equivalent to a single aperture 200-meter telescope that would cover the whole area atop the redesigned Cerro Paranal mountain.

Although the first interferometric fringes will be observed with two of the 8.2-meter units, eventually, the auxiliary telescopes will do detection of Jupiter-mass planets out to 1,000 parsecs at an accuracy of 10 microarcseconds. (That's almost equivalent to the SIM spacecraft, which will detect Jupiter-mass planets that lie at great distances.) VLTI will be capable of

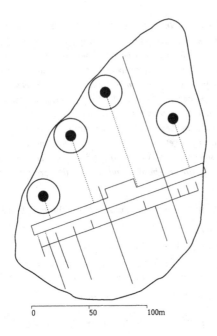

A schematic aerial view of the VLT. Each of the four Unit Telescopes are shown as large, filled circles, while solid lines show the location of tracks for transporting the smaller auxiliary telescopes between stations. The dotted lines indicate subterranean light ducts. (European Southern Observatory.)

detecting a Uranus-mass planet out to 200 parsecs, as well as planets ten times the mass of Earth around nearby stars. And the VLTI should also be capable of detecting a 1-Mj planet at 5 AU, the distance of our own Jupiter from the Sun.

The VLTI's Astronomical Multiple Beam Recombiner (AMBER) will link two of the 8.2-meter telescopes in an effort to do spectral analysis of massive "Jupiters" in close orbits around their planets, like 51 Pegasi b, the planet first discovered by Mayor and Queloz. Bruno Lopez, an astronomer at France's Observatoire de la Côte d'Azur near Nice, plans on using the VLTI and its state-of-the-art instrumentation to search for phase differences in the spectra of extra-solar planets in short orbits around their parent stars. Though it has nominal differences, Lopez's method will be very similar to that being planned by the Caltech team for use at Hawaii's Keck Observatory. According to Lopez, "A planet will provide a variation in the fringe contrast, which is going to be very small. This is why direct detection is very difficult. But if we accurately measure the fringe contrast at the various wavelengths, we can constrain the extra-solar planet's atmospheric spectrum. What we want to measure is the differential phase between two wavelengths: the distance between the fringes made at two different wave-lengths."[4]

By using two interferometrically linked 8.2-meter Unit Telescopes at the VLT, Lopez believes he could detect a hot "Jupiter" like 51 Pegasi b in two hours or less. Using the smaller telescopes, the same detection might take 100 hours. "A year after the first VLTI observation we [will be able to] tell if we have a good chance of observing extra-solar planets," says Lopez. "We know that we can observe one or two nights with the big telescopes, but if we demonstrate that

A view along the VLT's Interferometric Tunnel, with the rails for two of the VLTI Delay Lines. This interferometric capability will play a major role in the VLT's planet-hunting efforts via astrometry. The photo was taken on September 10, 2000. (European Southern Observatory.)

the measurements can be done, then I think we would get more time. If we jump on the telescope and fail, it will not [bode well] for getting additional observing time on the larger 8.2-meter units."

VLT OBSERVING TIME

The VLT is an ESO-owned and -operated installation, meaning it gives preference to astronomers affiliated with its member countries and to Chilean astronomers (up to 10 percent of VLT observing time) who have what ESO deems "meritorious" observing proposals. That restriction doesn't affect two Doppler spectroscopy veterans, Martin Kürster (Germany) and Michel Mayor (Switzerland), as both work for institutions from ESO-member countries. The two have both independently expressed interest in using the new VLT for planet hunting. Mayor told me that he will focus on Doppler spectroscopy in the next five years, but could see himself moving into planet-hunting via astrometry if the VLTI holds promise.[5]

ANTU SETS, YEPUN RISES

By late afternoon, the jackhammers were silent, and the night crew had begun chasing everyone but astronomers off the mountain. But Jorge Ianiszewski, ESO's VLT press liaison, knew a nice sunset was in the offing and was determined that I get a good view. So we headed partway down the mountain to a perch with a clear view of the Pacific, which is only 12 kilometers away. At this distance, the colorful aquamarine of the ocean looked like a veritable biological soup in comparison with the landmass on which we were standing. Condors are known to swoop over the telescopes every now and again, but seagulls don't venture this far; and aside from two types of foxes and a handful of small birds, there isn't much else here in the way of wildlife. The sky was wild enough. It was the beginning of a perfect evening: the seeing was so good that Venus was visible nearly a full half-hour before the Sun finished its descent into the

Pacific. The colors had already begun their transformation into a deeper hue of blue, and as the temperature dropped and the Sun made a quick exit, I heard Antu (the 8.2-meter UT1 telescope) rotating to life.

A few hours later, in the parking lot of the base camp, it was a real-life planetarium out there, more jolting than any *Star Wars* episode. A quarter moon had just bid us farewell, as Jorge set up his small 11.5-centimeter Celestron telescope. I pulled on the wooly black-billed cap that I had found wandering around Antofagasta's street market late the previous afternoon. Its ear flaps made me look like a wayward duck hunter, but it kept my ears from freezing in this wintry desert wind. Here I was on the fringes of "big-time" astronomy shamelessly looking through a backyard-quality telescope. In the end, fighting cold and fatigue, I finally took leave of Mars, Alpha Centauri, and several star clusters for a few hours sleep. The next afternoon, it was shirt-sleeve weather again, and the sky was its normal bluish-black. Riding back across the desert to Antofagasta, the Sun and the bumpy road were both just as relentless as the day before. And there was not a hint of what can be seen when the Sun goes down out in this no-man's land. After a night on an observatory mountain, E-mails left unanswered and faxes gone missing seem trivial indeed. I knew that a thousand miles south and several thousand miles north, the rest of the world was racing hither and yon. But for me, at that moment, my own existence seemed as precarious as the shallow atmosphere separating us from the nothingness beyond. If only for a fleeting minute, it was impossible not to imagine Earth as it really is: an oasis in the midst of a relative void.

1   Spyromilio, Jason, astronomer and head of VLT commissioning. Interviewed on July 19, 2000, at Cerro Paranal, Chile.

2   Quattri, Marco, ESO Project Manager for the VLT's Structural Mechanics. Interviewed on July 19, 1999, at Cerro Paranal, Chile.

3   Jaffe, Walter, staff astronomer, Leiden Observatory, the Netherlands. Interviewed on July 20, 1999, at Cerro Paranal, Chile.

4   Lopez, Bruno, astronomer, Observatoire de la Côte d'Azur, Nice, France. Interviewed on October 26, 1999.

5   Kürster, Martin, astronomer, ESO La Silla, Chile. Interviewed on August 5, 1999, at Bioastronomy 99, Hawaii.

Mayor, Michel, astronomer, Geneva Observatory. Interviewed on May 6, 1999, at Observatoire de Haute-Provence, France. A follow-up took place on September 8, 2000.

The Scope of Things to Come
CHAPTER 14

By the time I hit the futuristic control room of Mauna Kea's Gemini North Observatory, my head was spinning, but not because of any rapid-fire revelations I was having about the wonders of the Universe. It had to be the altitude. The place, a dormant volcano-cum-astronomical science reserve atop Mauna Kea on the big island of Hawaii, was abuzz with the mid-afternoon activity of a busy construction crew banging about in heavy boots and hard hats. I took a sip from my water bottle and tried to stay alert, but that wasn't easy at more than 4,200 meters (almost 14,000 feet). I had just begun to catch my breath when a burly man with a penetrating gaze thrust a clipboard holding a 12-page document into my hand. Dave Logan, the Gemini North Observatory's company clerk, didn't want me to read this bureaucratic treatise word for word. He only wanted me to initial it as fast I could, so that he could get back to juggling phone calls coming into the ground-floor office of this sleek new telescope. "I got six calls at once, which is pretty hard to do with only four lines," says Logan, passing me a pen. "These are not complimentary pens, by the way." Then he launched into his cautionary spiel,

In the control room of the Gemini North Observatory atop Mauna Kea during the author's visit in August 1999. From left to right: Peter Michaud and Dave Logan. (Photo by Bruce Dorminey.)

which he deadpanned in a rapid-fire staccato. (It was hard to know whether to laugh or get up and run down the mountain as fast as I was able.) "First page," Logan began, "You may suffer headaches, tiredness, irritability, inability to talk, anorexia, insomnia, reduced intellectual capacity, and vomiting. We also have a small percentage [of people] who black out cold and fall flat on their face. It's also possible to develop one of the more severe mountain sicknesses, such as pulmonary or cerebral edema, both of which can be fatal, which means as soon as you get here, the brain explodes and your lungs explode. But since you are not exploding, you don't have that. But if you die while you are here, it's your fault because we already warned you. If here longer than two hours, people tend to forget things like their address and their mother's maiden name because of lack of oxygen. This affects your judgment. One hour of unprotected sunlight at this altitude can take away your night vision for two days. So we all wear shades as soon as we step out. But we haven't seen the Sun since yesterday afternoon. And since we are still under construction, you will encounter hazardous situations. We do have our own ambulance on site, but it's an hour and a half drive down to Hilo medical center."[1]

By the time Logan had finished his "motivational" speech, I could hardly move. When I did get up I dropped everything I was holding: tape recorders, hat, backpack, batteries, microcassettes, Gemini brochures, and water bottles. I felt as if I might pass out in a fit of altitude sickness, a victim of astronomical over-zealousness. Therefore, our tour was brief and to the point—up 15 floors to Gemini' North's silver dome (the highest point in the Pacific), then back down again. That was about all I could stand. Clearly, the hour I had spent earlier acclimating at Hale Pohaku at the Onizuka Visitor's Center, the Mauna Kea reserve's midlevel facilities (named for the Hawaiian astronaut killed in the 1986 Challenger accident), hadn't been enough.

This exterior view of Mauna Kea highlights the Gemini North Observatory, which is the largest dome near the center of the image. This photo was taken in early 1999 shortly after a significant snowfall, a reminder that the caldera of this dormant volcano is home to Poli'ahu, the Hawaiian snow goddess. (Photo courtesy of Gemini Observatory.)

Surprisingly, some astronomers are known to forgo this midlevel stop altogether. But at Gemini North's dedication ceremony, which included Britain's Prince Andrew and the governor of Hawaii, about half of the visiting dignitaries were on oxygen. In hindsight, I should have been, too.

GEMINI'S GENESIS

None of the altitude-defying staff of Gemini North would have been there at all if not for English astronomer Sir William Herschel. Two centuries earlier, in December 1800, Herschel was conducting an experiment to measure the "heating powers of colored rays," specifically from the colored rays of sunlight as it emerged from a prism in its dispersed form. He noticed a temperature increase as he moved his thermometers from the violet to the red part of the "rainbow," and, to his surprise, recorded the highest temperature just beyond the red end of spectrum, where there was no visible light. Herschel surmised that this heat must be caused by unseen radiation, which we now term infrared.

Infrared (IR) radiation, emitted by everything from stars and protoplanetary disks to men and machines here on Earth, has been increasingly used as a window onto celestial bodies since Herschel's initial discovery. By studying IR, astronomers have been able to study many IR-emitting astronomical sources that might otherwise be obscured by cosmic dust.

The infrared spectrum began to fully open to ground-based observations in the early 1960s after the development of new filters and detectors that gave observers both better IR sensitivity and the opportunity to screen out IR radiation emitted by Earth-bound objects, including the telescopes' own instrumentation. Yet another challenge remained. Because water vapor and carbon dioxide (CO_2) in our own atmosphere block most incoming IR radiation from space, astronomers were forced to find a way to pierce this veil if they were to maximize their observations. Enter Mauna Kea, whose summit is already 40 percent of the way through our atmosphere and is arguably the best site for ground-based infrared observations. The fact that the big island of Hawaii is unattached to any continental landmass, some 2,200 kilometers north of the Equator, makes the air in the local area very stable. And with only 10 percent humidity near its peak, 75 percent of Mauna Kea's nights are good for seeing.

So, in 1987, when a U.S.-led international consortium, the Gemini Observatory Project, decided to build two large optical/infrared telescopes—one in the Northern Hemisphere and one in the Southern—it's no surprise that they picked Hawaii's Mauna Kea as one of the sites. (The other was Chile's Cerro Pachon, a 2,715-meter high mountain 10 kilometers southeast of the CTIO Blanco telescope). The two locations provide the $184-million Gemini Observatory Project full-sky coverage for their 8.1-meter mirrors. According to the terms under which the consortium was founded, the U.S. controls half of the observing time; the U.K. gets 25 percent; Canada, 15 percent; and Argentina, Chile, and Brazil divide the rest. Both observatories are open to the member countries' astronomical community at large, with observing time allotted based solely on the merits of each applicant's observing proposals.

Because Gemini North's viewing conditions are truly world-class, astronomers continue to salivate over the prospects of bigger and better astronomical instruments atop this rust-colored landscape. However, the volcano and its calderas have a very different significance to some indigenous Hawaiians, many of whom believe that Mauna Kea is the home of Poli'ahu, the Hawaiian snow goddess. According to legend, Poli'ahu's sister and bitter enemy, Pele, the volcano goddess, lived on Mauna Loa (Mauna Kea's volcanically active counterpart), while Poli'ahu kept her distance on Mauna Kea. Today, Poli'ahu shares her home with 13 telescopes, all perched inside or along the old volcanic rim. Many native Hawaiians would rather preserve their cultural heritage at this sacred site. Because of this conflict, all astronomical construction atop Mauna Kea must have local government approval.

Until changes in the local political climate permit construction of new astronomical observatories atop Mauna Kea, existing telescopes there will continue to play an important role in planet-hunting. In the realm of the search for extra-solar planets, Gemini North will specialize in Doppler spectroscopy surveys as well as astrometry and will be capable of spotting brown dwarfs and substellar-mass companions within 100 parsecs of Earth. It also comes equipped with a coronagraphic imager that should enable it to directly image Jupiter-mass planets in short orbits around their parent stars. According to Matt Mountain, Gemini Observatory's director, the North's infrared sensitivity will be ten times that of CTIO's 4-meter Blanco telescope.[2] In theory, Gemini North could spot a pair of headlights in San Francisco from its vantage point in Hawaii and should be able to do spectroscopic detection of a full range of chemical species within forming planetary systems.

From its location at 30 degrees south latitude, Gemini South will have a straight-on view of the galactic center, and when astronomers begin planet-hunting operations there in 2002, they will do high-precision Doppler spectroscopy. Meanwhile, the University of Florida's OSCIR (Observatory Spectrometer and Camera for the Infrared) instrument will initially be

This 360-degree panorama of the interior of Gemini North was taken from a catwalk on the inside of the dome, whose top marks the highest point in the Pacific. The telescope's 8.1-meter primary mirror coupled with a coronagraphic imager provides an ideal venue for infrared observations, possibly the direct detection of so-called hot Jupiters in short orbits around their parent stars. (Photo courtesy of Gemini Observatory.)

used on Gemini North and then be loaned to Gemini South. At this writing, the University of Florida group is building yet a higher-sensitivity infrared imager called T-ReCS, for Thermal Region Camera and Spectrometer, that will eventually share its time between the two Geminis and should be installed on Gemini South by November 2001.

Sea-Level Research

Astronomers who are susceptible to altitude problems will be spared the heavy breathing at both Gemini locations thanks to their remote operations capabilities at sea-level control centers in Hilo, Hawaii, and La Serena, Chile. In Hawaii, most of the work will take place at a new sea-level, 5,200-square-meter Gemini headquarters, the hub for Gemini's worldwide communications network. In some cases, astronomers can simply apply for time, outline the surveys or the objects they wish to target, and wait for it to be done by Gemini staff. The staff can thus ensure that the observations are done on a first-come-first-served basis—and at the optimum observing times. When the observations are not particularly complicated, astronomers might not have to visit at all; they will simply log on to the Internet to monitor what is happening at the observatory from their own time zone. But whether on-site or off, a night at either observatory is worth approximately $33,000, which is one reason two staff troubleshooters will always remain at each of the summits. And both Gemini North and South will play a special role in supplementing infrared observations now being done at Gemini North's older neighbors: the twin Keck I and II 10-meter telescopes.

EVOLUTION OF KECK

Marcy and Butler have made many of their Doppler spectroscopy discoveries using the Keck I, an eight-story, white-domed, 300-ton telescope that is only a short drive away and some 15 meters lower than Gemini North. But when Peter Michaud, Gemini North's press liaison, Jeremy Bass, a Gemini summer intern, and I pulled up to the Keck I late that same afternoon, it remained shrouded in clouds. We made a short trek through the volcanic muck that surrounds the telescope and entered through a side door leading to the control room. To our immediate left was the entrance to the Keck I's main mirror, where the lines that make up the 36 hexagonal segments of its primary mirror—which work as one single piece of reflective glass—are easily distinguishable, even within its darkened dome. At first glance, the segmented mirrors reminded me of the eye of a common housefly, as they lacked the perfectly smooth visual texture so evident in the monolithic 8.1-meter mirror at the nearby Gemini North. However, both are equally effective.

At the time of this writing, the William M. Keck Observatory was undergoing a $50-million NASA-funded upgrade to add four 1.8-meter telescopes and six interferometric delay lines to the observatory's $140-million twin 10-meter telescopes. Upon final completion at an unspecified later date, the upgrade will allow all six telescopes to be linked interferometrically. The result will give the Keck Observatory an optical range equal to that of one large 100-meter telescope, which is theoretically capable of spotting a housefly alighting on the surface of the Moon. And with the capability to do interferometric nulling, the project will also enable the direct detection of Jupiter-sized planets circling other stars, as well as give astronomers the chance to survey 500 nearby stars in search for planets as small as Uranus.[3]

The Global View

Keck generally operates in the infrared, and in 2002, it will begin using the Caltech team's method of differential astrometry for direct detection of Uranus-mass planets around as many as 1,000 nearby stars. The Keck Observatory will also measure local zodiacal dust in neighboring planetary systems and reexamine previously discovered extra-solar planets. While Keck and Gemini North are not in direct competition in the interferometric realm, each is active in the area of direct infrared planet imaging. Gemini North uses coronagraphy, while Keck uses interferometric nulling. Yet both claim superiority in their overall ability to do infrared ground-based imaging. "If you are working in the infrared, you are better off with Gemini," says Jason Spyromilio, himself an active infrared astronomer before he began his stint as head of commissioning at the VLT. "Keck's segmented mirror has heat sources all over its mirror, because all the gaps are actually hot. Gemini, with a monolithic mirror, is actually better in the infrared."[4] In other words, because the Keck has a segmented mirror, the gaps between its hexagonal segments are more prone to suffer from thermal pollution (causing higher temperatures, which can be detrimental to infrared telescopy) than an 8-meter-class mirror with a monolithic design. In any case, Marcy told me that he and his team had no plans to apply for time on the neighboring Gemini telescope, as he was perfectly happy with the results that they've been able to achieve with the Keck I.[5]

BEYOND GEMINI

Though it has detractors, the segmented mirror concept used at Keck I has been appropriated into a new line of low-cost telescopes, all of which will be searching for new planets. One, atop Mount Fowlkes in the far reaches of West Texas, has been operating since October 1999. The $15-million Hobby Eberly Telescope (HET) at McDonald Observatory was built for a fraction of the cost of comparable telescopes, but its 11-meter segmented mirror does not move. Instead, it's able to follow its targets as they move across the telescope's 91 1-meter segments through an overhead tracking focal assembly. Run by a five-university consortium (University of Texas at Austin, Penn State University, Stanford University, and two German universities), the HET uses a high-resolution spectrograph to do planet hunting via Doppler spectroscopy.

Taking a page from the HET, the South African Astronomical Observatory is building a similar structure on the Karoo plateau northeast of Cape Town. Dubbed the Southern African Large Telescope (SALT), this $16.5-million facility will be an international collaboration, which includes Polish, German, New Zealand, and American partners, and will be fully complete in 2005, when it will begin Doppler spectroscopy searches for planets.

Outside Observations

Before the end of 2002, a highly modified Boeing 747SP (a shortened version of the aircraft) called the Stratospheric Observatory for Infrared Astronomy, or SOFIA, will go into service as a 20-year replacement for the now-retired Kuiper Airborne Observatory. It will make its base at NASA's Ames Research Center in Moffett Field, California. The flying observatory's 2.5-meter mirror will be especially effective at 13,500 meters, where it will be able to take infrared observations above 99 percent of Earth's atmosphere. The telescope, designed and constructed by the German Aerospace Center, *Deutsches Luft- und Raumfahrtzentrum (DLR)*, in exchange for 20 percent of SOFIA's observing time, will be the largest airborne telescope in the world, but due to a special mirror lightweighting process will weigh only 20,000 kilograms.

On its four *sorties* per week, SOFIA will, it is hoped, be especially adept at taking high-resolution surveys of known protostellar planet-forming regions. Perhaps the next best thing to an airborne infrared telescope is an infrared telescope in space. But NASA's $1.2-billion Next-Generation Space Telescope (NGST) isn't currently scheduled for launch until 2009. With an extremely lightweight 8-meter mirror, NGST will be twice the size of Chile's 4-meter Blanco telescope, and 100 times more sensitive than the Hubble Space Telescope. Per square meter, it is to be 12 times lighter than Hubble's 1-meter mirror, making it easily deployable in orbit. If the NGST mirror is also fitted with a coronagraph, the telescope would be able to directly image a Jupiter analog at a distance of 8 parsecs from Earth within three hours or less. From its position at the second Lagrangian point (L2), it will also survey some 400 stars for exo-zodiacal dust, as well as for planet formation. But according to Peter Stockman, the NGST division head at the Space Telescope Science Institute in Baltimore, the NGST will still not be capable of

A graphic simulation of the European Southern Observatory's proposed 100-meter Overwhelming Large Telescope (OWL). Its segmented main mirror would be the largest ever constructed, and upon completion in the next decade the project would represent a revolution in both ground-based astronomy and planet-hunting. (European Southern Observatory.)

detecting Earth-like planets, for reasons that "boil down to cost." "NGST may have a simple coronagraph," Stockman adds, but notes that to have the capability to detect Earth-like planets would require technology for the NGST's primary mirror that would push costs up by a minimum of $50 million dollars.[6] No decision will be made on an NGST coronagraph until at least 2003.

A Ground-Based Revolution

Whatever happens with the NGST, and ESA's and NASA's other planned planet-hunting cousins, optical engineers predict that in another 20 years, telescopy from space will have lost its luster. With the advent of supertelescopes such as the Overwhelmingly Large Telescope (OWL), an estimated $800-million-dollar observatory now being considered by the European Southern Observatory, ground-based telescopy will be turned on its head. OWL would have a 100-meter main mirror composed of some 2,436 1.8-meter hexagonal segments. It would be almost as tall as the first level of the Eiffel Tower—some 25 stories. Its resolution would enable spectroscopy on extra-solar planetary atmospheres from the ground. It would also be able to pick off an extra-solar "Jupiter" at a distance of 10 parsecs within seconds and detect Earth-mass planets circling other stars some 30 parsecs away.

According to Philippe Dierickx, a European Southern Observatory optical engineer and OWL engineering team member, the idea for this monster was first broached in 1997. From full funding to final completion, OWL's realization would take 13 years—although it could begin observations earlier, once only several hundred of its mirror segments are in place. Funding would be the biggest obstacle to realizing such a gargantuan project. Out of necessity, Dierickx argues, the project would "very plausibly" become a worldwide collaboration. As to where the OWL might ultimately "perch," Dierickx says that ESO was actively considering several possible sites, including one in Chile, one near Uzbekistan's Maidanak Observatory in central Asia, and one atop Namibia's 2,400-meter high Gamsberg mountain in the Namib desert. No one site is absolutely ideal. However, the central Asian site offers low winds and little seismic activity, while Chile's Atacama desert is less appealing, as it might be too seismically active for such a technologically challenged instrument. In its favor, the Namibian L-shaped mountaintop already has a flat topography that would make it suitable for the installation of an astronomical observatory. Dierickx emphasizes that because at present there was no strong preference for any one site, a search for suitable sites is underway using the most recent satellite climate data.[7]

If OWL and other projects like it continue to proliferate, Mauna Kea's importance as an astronomical proving ground could gradually diminish and one day give indigenous Hawaiians the opportunity to see their dormant volcano reprise its singular role as home to the snow goddess Poli'ahu.

1   Logan, David, on-site construction manager, Gemini North Observatory. Interviewed on August 7, 1999, at Mauna Kea, Hawaii.

2   Mountain, Matt, Director Gemini Observatory. Interviewed on May 25, 2000.

3   Michaud, Peter, Gemini Outreach Director, Gemini North Observatory, Hilo, Hawaii. Interviewed on August 7, 1999, at Mauna Kea, Hawaii.

4   Spyromilio, Jason, astronomer and head of VLT commissioning. Interviewed on July 19, 2000, at Cerro Paranal, Chile.

5   Marcy, Geoffrey, astronomer, University of California at Berkeley. Interviewed on May 25, 1999, at Dana Point, California, and on August 6, 1999, at Hapuna Beach, Hawaii. Follow-ups took place on September 8, 2000, May 10–12, 2001, and June 3, 2001.

6   Stockman, Peter, NGST Division Head, Space Telescope Science Institute. Private communication with the author on May 24, 2000.

7   Dierickx, Philippe, ESO optical engineer, working engineering group for OWL. Interviewed on May 25, 2000.

Signatures of Life *CHAPTER 15*

On a side street in the back end of Paris's Latin Quarter, sandwiched between a boarding school and neighborhood post office, stands the building that once housed Louis Pasteur's lab. Here, in the latter half of the nineteenth century, the renowned chemist and microbiologist found a vaccine for rabies, developed the process that came to be known as pasteurization, and pondered a discovery that he suspected could have profound ramifications for the structure of life on Earth, if not the entire Universe.

In August 1847, Pasteur, a young doctoral candidate at the time, was trying to dissect the molecular structure of salts produced as a by-product of fermentation. (He and others had noticed that after fermentation, wine jugs and barrels often contained a residue of tartar, made up of potassium and tartaric acid.) Organic tartaric acid was found to rotate a beam of polarized light clockwise, whereas synthesized paratartaric acid showed no such optical activity. Using a dissecting needle and his microscope, Pasteur was determined to learn why.

TARTARIC AND PARATARTARIC ACIDS

Tartaric acid, a naturally occurring, dicarboxylic, optically active acid, is normally found in its white crystalline form in tartar (potassium hydrogen tartrate) deposits at the bottom of wine vats. First isolated in 1770 by Swedish chemist Carl W. Scheele, it is used in photography, medicine, dyeing and calico printing, and as a food additive. Paratartaric acid is chemically the same as its counterpart, tartaric acid, but is optically inactive, and thus does not rotate polarized light.

TAKING SIDES

Though both acids seemed to be chemically the same, Pasteur observed that the paratartrate was made up of pairs of left- and right-handed molecular crystals that were mirror images of each other. When he separated the two crystals, he found that one rotated plane-polarized light to the left (counterclockwise), and the other rotated the light to the right (clockwise). The Tartaric acid, however, was made up of only left-handed molecular crystals that rotated light to the right. To this day, it is still unclear why certain molecular compounds rotate light and others do not. Max Bernstein, an astrochemist at NASA Ames Research Center and the SETI Institute, points out that it is impossible to know which way light will rotate in any given compound. "We know that, based on molecular structure, you cannot predict which way the light will [rotate]," says Bernstein. "But it is consistent. In other words, if I measure the rotation of light from [the amino acid] L-alanine in my laboratory, it will rotate the light the same way in Australia or Alpha Centauri. But if you hadn't measured it before, you wouldn't know."[1]

Pasteur surmised that because the tartaric acid was a by-product of fermentation, itself a biological process, then its peculiar chirality (from *cheir*, ancient Greek for hand), must be a result of the "handedness" of life itself. Obviously, Pasteur couldn't know why or how life and handedness could be linked, for this is a conundrum that challenges us even today. But Pasteur remained intrigued by chirality's optical trickery and went further to speculate that understanding the origin of such asymmetry might also provide a key to the nature of life—specifically, that life was, in fact, a "function of the asymmetry of the Universe."[2]

Alanine (the smallest chiral amino acid) in its D and L forms in this schematic "stick and ball" molecular diagram. As an optical isomer, both D and L forms have the same chemical makeup, but they are mirror images of each other. In addition to being a crucial component of all living systems, alanine has also been detected in meteorites. (Based on a graphic by the Astrochemistry Laboratory at NASA's Ames Research Center.)

CHIRALITY

The vast majority of amino acids in all living things are based on proteins assembled from left-handed (L) amino acids, but these proteins metabolize sugars that predominantly exhibit right-handed (D) tendencies, denoted as L and D from the Latin leavus *and* dexter *(left and right). Compounds possessing this handedness, which sometimes enables them to rotate plane-polarized light, are called enantiomers or optical isomers. It is still not understood why L amino acids select D sugars, but the answer must lie in the Earth's early biochemistry. As NASA's Max Bernstein points out, in terms of evolution, all living things on this planet share the same biochemistry, in that we all use nucleic acids. "Nucleic acids give instructions on how to build proteins out of amino acids," says Bernstein. "We all learned this from a single common ancestor, and that common ancestor used L amino acids. We don't know why, but given that that's the case, imagine the evolutionary cost of rearranging your biochemistry so that you used D amino acids. That would be a big change."*

The Origin of Chirality

It may be that the origins of life's chiral nature could have less to do with asymmetry in the Universe as a whole and more to do with the influence of a nearby astrophysical event. A supernova, of the sort that could have triggered the collapse of the carbon-enriched gas cloud that resulted in the formation of our Solar System, may also have caused a left-handed molecular preference. Radiation from the newly formed neutron star at the core of the supernova may have caused the scattering of ultraviolet radiation in the dust particles as it passed through, causing the selection of one molecular hand over another. The *circular polarization* of light may in fact be continually influencing the handedness of molecules throughout star-forming regions of the galaxy, if not in the whole Universe. However, to appreciate how such handedness can come about in any given molecular structure, it is first necessary to come to terms with one of the natural astronomical processes by which light is thought to become circularly polarized.

CIRCULAR POLARIZATION

All photons of light propagate through space in one direction, but in unpolarized light, the wavelengths of the photons themselves vibrate in all directions in a manner perpendicular to the direction of propagation. In certain instances, whether intentional or by natural means, light can become plane-polarized after reflection, deflection, or transmission through some substances, so that its electric field vibrates in the same direction. Circularly polarized light is plane-polarized light that has been optically rotated so that its electric field moves in a helixlike three-dimensional curve, vectoring in both the same direction and at the same wavelength frequency. Elliptically polarized light, in contrast, is plane-polarized light that has been rotated so that its wavelength amplitudes form

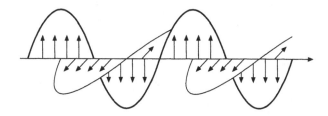

This graphic illustrates how circularly polarized light propagates along a horizontal axis. (Meadowlark Optics.)

right angles to the light's direction of propagation. Sunglasses normally contain a Polaroid material that plane-polarizes unpolarized light. Such sunglasses reduce glare by absorbing light that is vibrating in a horizontal direction, which would normally have been reflected off horizontal surfaces (such as highways).

Astronomers at the Anglo-Australian Observatory northwest of Sydney have observed circular polarization of infrared light spanning distances of 1,000 AU (many times larger than our Solar System) in the star-forming complex of NGC 6334, some 1,700 parsecs away from Earth in the constellation of Scorpio. Circular polarization can occur naturally through the scattering (deflection) of light as it travels through interstellar dust grains. Because the dust grains themselves may have already been aligned by magnetization, light passing through such magnetized grains may cause molecules that form in the local vicinity to form with a chiral preference.

Laboratory studies have confirmed that circularly polarized radiation can select (or destroy) one biomolecular hand in preference to another. Whether these lab studies are accurate representations of what happens in the interstellar medium has yet to be proven. However, we do know that strong magnetic fields are known to be common in star-forming regions, and that dust grains inside such magnetic fields tend to become magnetically aligned. So it follows that as ultraviolet radiation from a nearby supernova passes through this magnetic alignment of dust grains, some of the amino acids embedded in such grains (alanine, for instance) would also be destroyed, thus triggering the selection (or preference) of one biomolecular hand over another. "Light impinges on one [dust] grain," explains Bernstein, "and is more likely to destroy one hand of alanine over the other, so that when [the alanine] lands on Earth you have one preference over the other." A similar process may have happened in the star-forming region that spawned our own Solar System, since supernovae are thought to trigger the collapse of molecular clouds that form planetary systems. Thus radiation from the supernova which may have triggered the collapse of the molecular cloud that formed our own protostellar disk might also have had a local selection effect on amino acids, which eventually influenced the chiral preference of life on our own planet.

Although researchers have long known that chirality is requisite for Earth-based biota, they have also been eager to determine if chirality is indeed common in extraterrestrial compounds found in meteorites that have been collected by researchers here on Earth. Scientists frequently scrutinize meteorites to better understand their chemical composition. Thus for the astronomical community, it was oddly serendipitous that on September 28, 1969, the small town of Murchison in southeastern Australia suddenly found itself in the line of fire from an extraterrestrial intruder, in the form of a 4.5-billion-year-old meteorite. Most of the town's residents were reportedly in church that Sunday morning when a fireball exploded and hissed over them, scattering fragments of a meteorite (believed to be a cometary core) over an estimated 33 square kilometers. Some 700 kilograms of its charcoal-like fragments, including one weighing 7.5 kilograms, fell in and around the town. In 1970, researchers found that fragments from this carbonaceous chondrite contained organic molecules that included high concentrations of

L-amino acids, proving that some sort of asymmetrical process had begun long before life got started here on Earth.

"The fact that we see polarization in regions where stars are forming [and] where organic molecules are known to be present," says Jeremy Bailey, an astronomer at the Anglo-Australian Observatory, "tends to make a promising [point of origin] for the chiral symmetries that we've seen in Murchison. It also shows that chiral molecules can survive delivery onto Earth with their chiral properties intact. Our Solar System may have formed with an important prerequisite for life: left-handed amino acids." Bailey believes that if life's creation depends on asymmetric chirality originating in the chance polarization of organic molecules, then such conditions may be present only in certain parts of the galaxy, and thus may not be common throughout the galaxy or in other galaxies.[3]

Indeed, aside from the chemistry involved, it seems miraculous that life emerged at all. In the face of huge slabs of rock slamming into Earth's nascent surface, it's no wonder that life took several hundred million years to appear. Conditions on the surface of the young Earth were violently energetic; it was a place where all the sunblock in the world wouldn't be able to protect against UV radiation 40 times its present levels. There were hurricanes to end all hurricanes, and half-million-volt electrical storms. In spite of their hew and yaw and overall destructiveness, comets seeded the young Earth with water and most of the building blocks of life, including organics, such as amino acids and other carbon-based chemical species.

COMET
A comet is a small body made up of ice and dust in orbit around the Sun. Its nucleus is generally no more than 1 kilometer in diameter. As the comet approaches the Sun, it heats up, causing the release of gas and dust from its icy nucleus, which usually creates a coma (a teardrop-shaped gas and dust envelope behind the nucleus) and a tail. Comet tails have been known to extend up to 100 million kilometers, or about two-thirds of the distance between Earth and Sun. When near the Sun, the comet not only reflects sunlight, it also emits light from its ionized gases, which are heated by the Sun.

Given the age of the Solar System and the first appearance of microbial life on Earth, planetary scientists theorize that the heavy bombardment phase took place over a 700-million year epoch ranging from an estimated 4.5 to 3.8 billion years ago. Water is arguably the most important of all the elements that the bombardments are thought to have contributed to Earth. Many planetary scientists think that an overwhelming majority of Earth's initial ocean water came from this period's plethora of Earth-impacting comets. Earlier than 3.8 billion years ago, the uppermost layers of the ocean would have been repeatedly vaporized by large impacts.

But not everyone agrees that most of this water came from comets. Jonathan Lunine, a theoretical astrophysicist at the University of Arizona, believes that by the time Earth reached half its present mass, it was already being pounded by what he terms "water-bearing planet embryos." "As the Solar System formed," says Lunine, "Jupiter perturbed asteroids to accrete into larger and larger objects—terrestrial embryos as big as Mars or bigger—then tossed them

This montage of our Solar System is made up of photos taken by various spacecraft managed by the Jet Propulsion Laboratory in Pasadena, California. Included are (from top to bottom) images of Mercury, Venus, Earth (and Moon), Mars, Jupiter, Saturn, Uranus and Neptune. Pluto is not shown, as no spacecraft has yet visited it. (NASA/JPL/Caltech.)

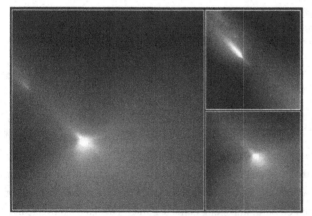

Comet Hyakutake, a long-period comet discovered in January 1996 by amateur Japanese astronomer Yuuji Hyakutake, was subsequently photographed by the Hubble Space Telescope's Wide Field Planetary Camera 2. This collection of images was taken on March 25, 1996, when the comet passed only 14.9 million kilometers from Earth, and shows the region near the heart of the comet's icy, solid nucleus. (NASA/JPL/Caltech.)

into very unstable orbits."[4] Lunine believes that these embryos carried enough water to provide more than three times the amount needed to make up Earth's oceans, and that much of the water these embryos delivered was also buried in Earth's crust and upper mantle. He credits comets with contributing no more than 10 percent of Earth's water, but much of its organic carbon. However, because primordial Earth was hit by a Mars-sized rock around 4.4 billion years ago, which in turn led to the formation of the Moon, most of Earth's primordial atmosphere, including water that had formed on the young Earth previous to the impact, would have been lost into space. According to Michael Mumma, cometary scientist at NASA Goddard Space Flight Center in Greenbelt, Maryland, this is where comets played their role, for they "reseeded" the Earth with both water and organics.[5]

The four giant outer planets—Jupiter, Saturn, Uranus, and Neptune—each acted as perturbing bodies scattering large masses of planetesimals, planet embryos, and comets in all directions. Both Mumma and Lunine agree that Jupiter kicked large numbers of objects into the inner Solar System, but they disagree on precisely which kind delivered Earth's water. Lunine believes they were water-bearing "planet embryos," while Mumma says that what he would classify as "comets" delivered the vast majority of Earth's water. "Jupiter contributed most of the comets that impacted Earth and the other terrestrial planets," argues Mumma. "They pelted the hell out of Earth. Jupiter ejected a lot of stuff of all sizes, from 10 kilometers in diameter to maybe even planetary embryos, on down to . . . meters in size." Mumma adds that the smaller objects were able to impact Earth and deliver most of their water and organics intact (without destroying their chemical integrity). The larger objects would have formed fireballs, but Mumma believes that much of these larger objects' water and organics would have survived impact in the protective cometary cores.

THE OORT CLOUD

Although an estimated 97 percent of the comets Jupiter perturbed were simply kicked out into interstellar space, the remaining 3 percent went into a comet reservoir located about 0.3 parsecs from the Sun, where they continue to circle to this day. Known as the Oort Cloud, this is the home of the long-period comets, including Halley, Hyakutake, and Hale-Bopp. Although this cloud of comets has never been directly observed, Dutch astronomer Jan Oort theorized in 1950 that it must be the source of a never-ending supply of new comets. (He based his assumptions on the sheer numbers of new comets that are continually observed approaching the inner Solar System from all angles and inclinations.) Each time a passing star comes within a short distance of our Solar System, this massive cloud of comets is perturbed, causing an effect akin to that of stirring up a nest of bees. When this happens, some of the comets eventually find their way into our neighborhood, the inner Solar System.

Algol, a multiple star system located an estimated 30 parsecs away from the Earth in Perseus, made a close approach to the Oort Cloud some 7.3 million years ago. The Oort Cloud's next closest stellar approach will be by Gliese 710, a reddish M or K star now less than 20 parsecs away. In "only" another million years, it will pass little more than 0.4 parsecs from the Oort Cloud. Using numerical models based on the amount of material believed to exist in the Oort Cloud, theorists estimate that the passing of Gliese 710 will cause perhaps as many as 2.4 million comets to move into new Earth-crossing orbits. These comets will arrive in our vicinity over a period of 2 million years. Some will circle repeatedly; others will have been so disturbed by Gliese 710's passing that they will be knocked out of the Solar System on their first lap around the interplanetary block. Regardless, Gliese 710 is certain to cause a 50-percent increase in the number of comets in Earth's path. Even so, astrophysicists say that Gliese 710's passing doesn't necessarily mean that Earth will automatically fall victim to cataclysmic cometary bombardments.

The most recent major asteroid to penetrate Earth's atmosphere may offer researchers the best chance ever of finding pristine compounds left over from the formation of the Solar System some 4.5 billion years ago. The asteroid hurtled across the northern Yukon in mid-January 2000, creating a huge fireball as it broke up in the atmosphere. Sonic booms and sizzling sounds were evident from Alaska to Northwestern Canada, and the unmistakable foul odor of sulfur was in the air. Eight days after it hit, an outdoor guide and pilot happened upon fragments of the meteorite as he was traveling across frozen Tagish Lake in his pickup truck. Anticipating such a windfall, researchers had already advised locals on how to retrieve the fragments, to ensure they weren't contaminated. Fortunately, the guide remembered the advice and collected the samples correctly. The following spring, a 13-member international research team returned, bringing with them all manner of tools, including chain saws, spoons, and chopsticks, which they used to retrieve as many of the carbonaceous chondrite fragments as possible before the ice melted and they sunk to the bottom of Tagish Lake.

ASTEROIDS
Asteroids are small rocky or metallic objects most often found in the so-called asteroid belt, the orbital region that lies between Mars and Jupiter. They typically range from 10 meters to 1,000 kilometers in diameter. More than a million are thought to exist, 95 percent of which lie in this "belt." It is believed that they are ancient building blocks of wanna-be planets that never got the opportunity to accrete into large bodies due to the disruptive effect of Jupiter's gravity. Some asteroids are thought to be fragments of cometary nuclei. Those that cross Earth's orbit are thus known as near-Earth asteroids.

A meteorite is the term given to any natural object that enters through and strikes Earth's surface, regardless of its initial mass when it enters our atmosphere. Thus, an asteroid that strikes Earth's surface is also automatically termed a large meteorite. A meteor is the visible light produced by a small meteoroid (a small object that has broken off from an asteroid or comet) as it plummets through Earth's atmosphere at high velocities and at high altitudes.

CHONDRITES
Chondrites, the most abundant class of meteorites, are truly the building blocks of the inner Solar System. Normally pitch-black, they vaguely resemble partially used briquettes of charcoal, rich in iron and magnesium-bearing silicates, and they make up almost 90 percent of the meteorites that fall to the Earth's surface. (See also "Chondrules" on page 29.)

The group found more than 500 fragments, the largest being somewhat bigger than a small potato. NASA estimates that the meteorite from which they came must have been at least 7 meters across, weighed up to 250 metric tons, and most likely originated in the main asteroid belt between Mars and Jupiter. The fragments from the Tagish Lake meteorite will be examined for decades to come for their organics, including carbon chain molecules, amino acids, and certainly their tendencies toward chirality.

THE WATER-CARBON CONNECTION

Without water, life as we know it here on Earth would never have emerged. So, of all the compounds that comets, asteroids, and meteorites are thought to have contributed to Earth, water may be the most critical. Water is essential as a solvent in the mixing and making of complex molecules and amino acids, which are necessary for the formation of ribonucleic acid (RNA), the precursor of deoxyribonucleic acid (DNA) and, therefore, life. It is now believed that after Earth finished its formation (accretion) process in the fray of the early inner Solar System, its atmosphere was mostly a hot mixture of dust and steam, which rained out into salty, dirty brown, iron-rich oceans.

But as central as water was to the formation of life on this planet, without carbon, Earth's early biology would have had a very rough time of replicating and diversifying. Carbon and water together gave life a durable structure, in which it could evolve and endure. *Star Trek* fans have long fantasized about silicon-based life forms (and, given the plethora of recent advances in silicon chip technology, it's not hard to understand why). But André Brack, a chemist and national researcher at France's Center for Molecular Biophysics in Orléans, says that carbon-based life is a safer bet. "I would not totally reject [the] silicon [theory], but with silicon, the chemistry would be much poorer, and the number of combinations are more limited. Silicon life would be life in insoluble form. It would not allow for fast diffusion of the components and would be very slow to reproduce."[6] The data bear him out: of the molecular compounds identified thus far in the interstellar medium (the mixture of dust, gas, and cosmic ray particles between the stars in our galaxy), 91 contain carbon, while only 8 contain silicon. Thus, if these current data represent the real carbon-silicon abundances in the Universe as a whole, then it seems likely that if life does evolve elsewhere, it would follow life's example here on Earth and structure itself around carbon.

Which Came First?

Precisely how life evolved out of the prebiotic soup of our early Earth remains a mystery, but it's a mystery for which we now have clues. That old saw about how life crawled out of a muck of protoplasm swimming in salt would seem to be true. Until 3.8 billion years ago, Earth was still being knocked silly by 440-kilometer-sized asteroids, which in turn vaporized what little ocean there was at the time, effectively sterilizing the planet. Deep seafloor hydrothermal vents were among the few places where bacterial life could withstand such impacts. These vents lie along fault lines in Earth's crust and churn out a hot mix of gaseous methane, carbon, and sulfur, upon which some bacteria seem to thrive.

Few scientists agree, however, that life started in these vents. "Even if life had its origin in hydrothermal vents—and that's a big if," says Francesco Santini, a zoologist at the University of Toronto, "certainly, most life has evolved outside of hydrothermal vents. The organisms that live there are organisms that have adapted to life [under such conditions]."[7] Wherever life first evolved, we do know that Earth was uninhabitable to any sort of life for at least a half billion years after it formed. That only leaves life with a 200- to 300-million-year window of opportunity in which to develop. And as everyone now knows, the key component of all life is DNA. (DNA is the key information processor of the biological world and contains all the details on how proteins are to be manufactured, simply by the order in which 20 different varieties of amino acids are arranged. Usually, one gene contains the recipe for making one protein. From there, proteins do much of the rest.) But which came first, RNA or DNA, is a variation on the old chicken-and-egg question. Most biophysicists now believe that RNA emerged from ocean sediments containing sulfur-rich clay compounds called thioesters. Chemical bonding in thioesters may have fueled a sort of prebiotic metabolism leading to the chains of proteins that form RNA. DNA then evolved from RNA.

The Sun's Role

However early life on Earth started, it is remarkable that it was able to sustain itself under the steady barrage of the Sun's UV radiation, which destroys DNA. Because of this, most biologists believe that life started in the oceans, whose salt, clay, and iron afforded some protection from the Sun's unrelenting output. In this early era, oxygen and ozone (which protects us from UV rays) was not present in sufficient quantities to afford an ozone layer. It wasn't until the advent of photosynthesis and biotically produced oxygen that life could tolerate the Sun's radiation and make its way on land.

Ironically, oxygen was poisonous to the single-celled precursors to modern-day bacteria, which were the first life forms to appear. As anaerobes, they couldn't abide the one element that we humans regard as the singularly most important to our survival. Our white blood cells still use what are known as free radicals to kill invading microbes. But free radicals—the ancient toxic derivatives of oxygen—also contribute to aging and cause cancer. Today, bacteria remain the world's most prolific life form. Each of us carries around about a trillion bacteria on our skin, 10 billion in our mouths, and 100 trillion in our gastrointestinal tracts.[8]

The emergence of cyanobacteria (blue-green algae) marked a true turning point in the development of life on Earth. They began using visible light as a means of acquiring additional carbon to continue the synthesis of more molecules. (Fossils of the earliest cyanobacteria, which are up to 3.8 billion years old, have been found in sedimentary columns in western Australia.) For years, theorists have debated the exact timing of the onslaught of oxygen; but it is now known that, perhaps as early as 2.7 billion years ago, this early photosynthesis resulted in the gradual yet significant manufacture of oxygen.[9]

Timing the appearance of ozone is also problematic. What is known is that Sun's UV radiation caused a massive ionization of oxygen that resulted in the protective ozonosphere under which we all live. Ozone formed only after the Sun's UV radiation had oxidized the ocean's ferrous iron and sulfide into forms that were insoluble, thus allowing cyanobacteria to move nearer to the surface. As cyanobacteria came to the surface, they also moved closer to Earth's burgeoning shores. Although cells remained of the single variety until 700 million years ago, the long march to land had begun almost 3 billion years earlier.[10]

In November 2000, researchers from NASA's Astrobiology Institute (NAI) reported that they had found fossilized remnants of cyanobacterial mats that may have developed on land as early as 2.6 billion to 2.7 billion years ago. The remnants were taken from samples collected in a 55-foot-thick layer of soil in Mpumalanga Province, in the Eastern Transvaal district of South Africa. The researchers theorize that the cyanobacteria could have developed in the clay-rich soil during the region's rainy seasons. (Previously, the oldest-known land-based microfossils dated back to 1.2 billion-year old microfossils, which were found in 1994 in Arizona.)

LIFE IN SPACE

Microbial life's extraordinary tenacity for establishing itself in the extreme environments of the young Earth makes one wonder whether it could develop and survive in space. Jay Melosh, a planetary scientist at the University of Arizona in Tucson and a member of an international team investigating such transfer scenarios, contends that the interstellar transfer of microbial life is highly improbable, even for distances to Proxima Centauri, our nearest stellar neighbor.[11]

However, we do know that microbial life can survive in the vacuum of the lunar surface for extended periods. Proof came shortly after NASA's Apollo 12 mission to the Moon. In November 1969, the lunar module landed on the Ocean of Storms (Oceanus Procellarum), some 1,500 kilometers west of Apollo 11's Tranquility Base. There, astronauts Pete Conrad and Alan Bean sauntered down a crater to pay a visit to an old friend: Surveyor 3, NASA's first probe to successfully land and transmit pictures from the lunar surface, which had arrived on the Moon almost three years earlier. Using a bolt cutter, the astronauts extracted the Surveyor's TV camera and its scoop for the return trip to Earth. Upon the camera's return, NASA successfully cultured spores of Streptococcus mitis, a common, harmless, human bacterium, which they had extracted from a strip of polyurethane foam insulation inside the camera. Some 100 of the single-cell organisms had survived the Surveyor's launch, the trip through the vacuum of space, the lunar landing, almost three years of exposure to radiation, and a veritable deep freeze. With no nutrients, water, or energy source, Streptococcus mitis had formed spores, which survived and were then recultured in a bacterial medium back on Earth. (The Centers for Disease Control and Prevention in Atlanta confirmed NASA's finding that, indeed, the bacteria were alive and well.) However, according to André Brack, 300 million years of prebiotic chemistry would probably be enough to usher in life on Earth, without the need for any external microorganisms arriving here after an eons-long amble across the local interstellar medium. In Brack's view, given the right circumstances, life will arise and doesn't necessarily need to be imported from space.

But whatever the dynamics of the origins of life, before we go looking for it elsewhere, we need to be certain that we *can* in fact identify life from space. This was the goal of one of the late Carl Sagan's most famous experiments. In 1990, he lead a team that used NASA's Galileo spacecraft and atmospheric probe to prove that life here on Earth could be detected from space, using currently available spacecraft instrumentation. The logic of the experiment was, if our spacecraft cannot collect objective evidence of life on Earth, how would we ever be able to detect life's signature elsewhere? Sagan's team took advantage of Galileo's early Venus and Earth flybys, where the craft gained acceleration through gravity-assist trajectories and thus continued on its long journey to Jupiter and its moons. In December 1990, 960 kilometers above the Caribbean Sea, the spacecraft's sensors detected abundant gaseous oxygen, ozone levels more than 20 times normal for a planet with no life, and methane at levels of one part per million, about 140 times more than expected without some underlying biological activity. Thus, the spacecraft provided the team with hard evidence that Earth, at the very least, presented evidence for simple vegetation. This sort of test may seem bizarre, but of course what Sagan and his team were testing was not the likelihood of life on Earth, but the accuracy of the technology that may some day enable us to spectroscopically detect life on extra-solar planets.

Imaging Life

Sensing life on close passes to a planet where we are certain there is life is one thing. Detecting life via spectroscopy in an atmosphere several parsecs away is quite another, presenting astronomers and the burgeoning field of astrobiology with what may be their greatest challenge. NASA is currently exploring the possible use of the Digital Array Scanned Interferometer (DASI), a digital imager that has been used for agricultural surveillance on Earth, for a possible future NASA flyby of Pluto. DASI would provide the space agency with its first look back at Earth at such a distance. However, to the imager Earth would likely look like a planet around just another star—a perspective that would give researchers an idea of what it would take to obtain spectra from other Earth-like planets around nearby stars.

If DASI makes it possible to remotely detect life on Earth from the outer reaches of our Solar System, it might also be possible to spectroscopically detect life in other planetary systems. ESA is studying the idea of using remote observations from spacecraft to detect homochirality. The researchers will first use this method to survey for the homochiral relics of past or present microbial life on Mars, then in the coma of a comet, and finally from the reflected light of extra-solar "earths." A planet with Amazon rainforest-like vegetation, for example, would have leaves or leaf-like structures that might reflect light bounced off chiral chlorophyll-like molecules. The reflected light, in turn, would produce a miniscule amount of circular polarization, which might be detectable some 20 years from now using future space-based interferometers (which both NASA and ESA are proposing).

ESA would like to put an instrument on its Mars Express mission, due for launch in 2003, that would remotely detect homochirality in the relics of what might be the red planet's now-extinct microbial life. Researchers estimate that small pools of the homochiral signature of life might still be detected even if life on the planet went extinct 3 to 4 billion years ago. How is this possible? The homochiral molecules rotate polarized light in opposite directions, so the instrument would measure the angle of the resulting optical rotation with a polarimeter and thus remotely measure a planet's homochirality.

Also in 2003, ESA may place a similar homochirality detector on its Rosetta mission to the comet Wirtanen. Instead of using a polarimeter, Rosetta would use gas chromatography, a method of separation and analysis, to divide cometary material into L- and D-handed molecules as it passes through a column inside the spacecraft. By determining the relative amounts of each type of L- or D-handed molecule, astronomers could make assumptions about the material's chiral orientation and thus the prevalence of left- or right-handed molecules in our Solar System.

In the more distant future, a second generation ESA IRSI/Darwin mission, now talked about for launch between 2020 and 2025, might finally be able to detect homochirality on extra-solar Earth-like planets around nearby stars. The appearance of a global-bulk homochiral signature in the spectrometers of such future missions would provide strong evidence that such an extra-solar planet must harbor some sort of active biology. Needless to say, any results from such missions would again be the subject of much debate at conferences of the future, for they would inherently be subject to interpretation and ambiguity.

Some present-day astronomers, Antoine Labeyrie for one, would like to see ESA or NASA opt for the actual extra-solar imaging of photosynthesis. To that end, Labeyrie has proposed the Exo-Earth Imager (EEI), an interferometer consisting of 150 separate 3-meter space telescopes arrayed in concentric circles over 100 kilometers. They would have the combined capability to detect green spots similar to Earth's Amazon basin on extra-solar planets that are 3 parsecs from Earth, and to do imaging in the visible or infrared spectrum. NASA does, in fact, have a project comparable to Labeyrie's on the drawing board, though its implementation is not on the "immediate horizon." Its Planet Imager would consist of five interferometers flying in formation, each carrying four 8-meter telescopes that would collect and combine their light with the help of a separate 8-meter telescope over a distance of some 6,000 kilometers.

The idea for NASA's Planet Imager was submitted in 1995, after Dan Goldin, the agency's administrator, asked what it would take to get pictures of continents and landmasses on extra-solar "Earths." Roger Angel from the University of Arizona is not sanguine about the feasibility of such planet imagers in the near future: "It would take several magnitudes more telescope than is needed to do spectroscopy. It's not lunacy to think about how you want to see this technology evolve, but it's one thing to write down the physics of what's required, and another to actually implement it. The laws of physics tell you what's needed but they don't tell you how to make it."[12]

Still, we can take heart in the fact that current technology is taking us to both ends of the spectrum. We are now going through the long, painstaking, and sometimes agonizing process of discovering exactly how our Solar System formed, evolved, and thus developed life here on Earth. At the same time, we are just beginning our quest for discovering life outside the Solar System, and in the process, we are honing our definitions of what constitutes life on our own planet. Certainly, this new, hard-won knowledge of our biogenesis will reap dividends down the line, as we become more aware of exactly what signatures of life we should heed.

Ironically, at the beginning of this new millennium, we appear to be as fascinated by finding microfossils on the shores of ancient Africa as we are in searching for life around stars parsecs away. No matter the outcome, our parallel quests for life both high and low is now giving us insights into both ends of the evolutionary continuum.

1   Bernstein, Max, astrochemist, NASA Ames Research Center and the SETI Institute. Interviewed on July 5, 2001.

2   "La vie est fonction de la dissymétrie de l'Univers. . . ." From "Louis Pasteur," Service Scientifique de l'Ambassade de France au Canada website, http://ambafrance-ca.org/HYPERLAB/PEOPLE/pasteur.html.

3   Bailey, Jeremy, astronomer at Anglo-Australian Observatory. Interviewed on August 7, 1999, at Bioastronomy 99, Hawaii.

4   Lunine, Jonathan, theoretical astrophysicist, University of Arizona at Tucson. Interviewed on March 29, 2001.

5   Mumma, Michael, cometary scientist, NASA Goddard Space Flight Center, Greenbelt, Maryland. Interviewed on April 5, 2001.

6   Brack, André, chemist and national researcher, Conseil National des Recherches Scientifiques (CNRS), Orléans, France. Interviewed on June 5, 2000.

7   Santini, Francesco, zoologist, University of Toronto. Interviewed on August 4, 1999, at Bioastronomy 99, Hawaii.

8   de Duve, Christian René 1995. *Vital Dust: Life as a Cosmic Imperative.* New York: Basic Books: 136, 126.

9   Barrow, John D. and Frank J. Tipler 1986. *The Anthropic Cosmological Principle.* Oxford, U.K. & New York: Oxford University Press: 548.

10   de Duve, Christian René 1995. *Vital Dust: Life as a Cosmic Imperative.* New York: Basic Books: 135.

11   Melosh, Jay, planetary scientist, Lunar and Planetary Laboratory, University of Arizona. Interviewed on June 6, 2000.

12   Angel, Roger, astronomer, University of Arizona at Tucson. Interviewed on June 1, 1999.

CHAPTER 16

Evolution, from the simplest multi-cellular life to hominids roaming the savannas of East Africa to our present space-faring civilization, all happened within 5 billion years of our planet's formation. Whether Earth would see the same result if we could wind back and replay the evolutionary tape of life, as paleontologist Stephen Jay Gould has wondered, remains very much an open question.[1] "To me the most profound and puzzling question," says Geoffrey Marcy, "is whether or not Darwinian evolution, which certainly leads to survival among the life forms, necessarily vectors toward intelligence, dexterity, and communicative skills. It's not clear that if you started another Earth-like planet with the same conditions that you would vector toward intelligence. You might end up with a lot of cockroaches and woodpeckers, and maybe a few whales, but it's not clear that the galaxy is teeming with intelligent life."[2] And yet we doggedly search for it.

CARBON COPIES

To humans, the idea of being alone in a Universe at least 10 to 13 billion light years across is disconcerting. The philosophical ramifications of being alone in such an overwhelming expanse of spacetime cannot be overstated. Given everything we know about the current rate of star and planetary formation, however, it would seem illogical that we should be the only sentient beings in the Universe. And given what we do know about the structure of life here on Earth and of molecules found thus far in the interstellar medium, it would appear that carbon would be the key to life's development anywhere in the Universe.

While we can only continue to speculate about the extraterrestrial development of life and intelligence, we do know that carbon production is based on the rate of star formation. Estimates by the Hubble Space Telescope strongly suggest that carbon production peaked almost 7 billion years ago. Given the time span necessary for biological evolution as we know it, some theorists now believe that it is highly unlikely that the Universe could have seen the first carbon-based intelligent life any sooner than 3 billion years ago, or when the Universe was already more than 10 billion years old. In other words, the evolution of extraterrestrial intelligent life could be a very "recent" cosmic phenomenon. In fact, as Mario Livio from the Space Telescope Science Institute and Charles Lineweaver from the University of New South Wales in Sydney have both pointed out, the Universe may only just be awakening to an epoch of intelligent life. As Lineweaver has noted, at this stage in our own development, it is impossible to know whether we have come late or early to the cosmic party, but in the long history of our Universe, we might be relative newcomers. This doesn't mean that other extraterrestrial civilizations might not have preceded us, but simply that extraterrestrial life, if it does exist, may be developing along cosmic time frames that are somewhat similar to our own. If such intelligences have evolved or are evolving in relative parallel to our own, would they be able to endure long enough as civilizations to allow for their probable detection over interstellar distances? As was poignantly illustrated in the Cold War, the answer may depend on how they will have marshalled their technology. Because we, ourselves, are such a new technology, these are questions that will take millenia to answer. Yet even if the time frames for the possible evolution of technologically-adept communicating civilizations remain fuzzy, there do appear to be some ground rules for how these communicating beings might evolve.[3]

Tuning in to ET

If there are extraterrestrial civilizations capable of communicating as we do, shouldn't it follow that the same basic physics also held for their evolution? Life as we know it has its best chance of developing on an Earth-like, fast-rotating planet in orbit around a Sun-like F, G, or K star, which has a main-sequence (that is, hydrogen-burning) phase that would provide a stable environment for life to evolve. And even with those parameters in place, ETs would likely emerge only after their home world had developed some sort of genetic code—if not RNA and DNA, then some form of biological information-processing system to speed the evolutionary plow. They would also have to become cognizant enough to communicate with each other.

Yet in order to communicate over interstellar distances, extraterrestrials would first have to overcome the gravity of their own planet. In order to communicate with us, as Seth Shostak of the SETI Institute frequently points out, they would have to develop the dexterity to build technology and telescopes. If they chose radio, they could easily overcome the universal cacophony of background noise in the electromagnetic microwave region (1,000 to 100,000 megahertz). Just above the terrestrial TV and FM bands, it's relatively quiet. There, only a trace of background radiation from the Big Bang remains. (This is an effect anyone can hear in the form of soft constant static when flipping the dials on an FM radio.) These quiet frequencies include what's been termed the "water hole," lying between 1,420 and 1,720 megahertz around natural emissions of hydrogen (H) and hydroxyl (OH), which in combination form H_2O. (All natural radio emissions come in the form of broadband, while all artificial radio emissions are narrowband emissions and come in continuous wave or pulsed signals.) So, astronomers interested in finding extraterrestrial intelligence in the radio spectrum are obviously most interested in these narrowband radio emissions. However, the pursuit of ET in the radio spectrum has a short history. Although decades, even centuries earlier, there were many ideas about how to signal or find extraterrestrial civilizations, the genesis of modern SETI really began only some 40 years ago.

In a paper published in *Nature* in 1959, the Cornell University physicists Giuseppe Cocconi and Philip Morrison suggested that the most effective way of communicating across galactic distances had to be via radio waves.[4] At about the same time, Frank Drake, a young radio astronomer at the National Radio Astronomy Observatory in Green Bank, West Virginia, was working on a pet project named Ozma. (The name came from the mythical Princess Ozma, featured in L. Frank Baum's fabled land of Oz: a place far, far away, populated with strange and exotic beings.)

Drake became the first radio astronomer to attempt to detect interstellar radio transmissions from an extraterrestrial intelligence. On April 8, 1960, he used the novel combination of sensitive new receivers and a 26-meter radio telescope to survey several nearby stars, the first of which was Tau Ceti, an early-morning star in the Cetus constellation some 3.3 parsecs from Earth. Drake picked up a strong signal almost as soon as he began scanning with his single, 100-hertz receiver at the 21-centimeter emission line (1,420 megahertz), which is the emission frequency of cold hydrogen from interstellar space. Yet some weeks later, Drake learned that the "signal" was really terrestrial interference from a secret U.S. military project.

Setting SETI's Parameters

Despite this frustrating experience with radio frequency interference (RFI), Drake's efforts prompted a request from the National Academy of Sciences asking that he organize a 1961 meeting to discuss the budding Search for Extraterrestrial Intelligence, or SETI. At this meet-

Frank Drake at the National Radio Astronomy Observatory's 91-meter telescope in Green Bank, West Virginia, about a decade after Project Ozma, Drake's first radio search for evidence of extraterrestrial intelligence. (NRAO.)

ing, Drake presented his now famous equation for determining the number of civilizations in our galaxy whose radio emissions are detectable. Now a professor of research astronomy at the University of California in Santa Cruz, Drake is also chairman of the SETI Institute's board of trustees. The institute is the largest scientific, privately funded, nonprofit research organization devoted to defining the numbers that make up each factor of his equation.

DRAKE EQUATION

$$N = R_* \times f_p \times n_e \times f_l \times f_i \times f_c \times L$$

Where,

N = the number of civilizations in the Milky Way galaxy whose radio emissions are detectable;

R_* = rate of formation of stars suitable for the development of intelligent life;

f_p = fraction of those stars with planetary systems;

n_e = number of "earths" per planetary system with an environment suitable for life;

f_l = fraction of those planets where life actually develops;

f_i = fraction of life-bearing planets on which intelligent life emerges;

f_c = fraction of civilizations that develop a technology which releases detectable signs of their existence into space;

L = length of time such civilizations release detectable signals into space.

Today, radio searches funded by the SETI Institute use equipment that is 100 trillion times more powerful than what Drake used on Project Ozma. One of the SETI Institute's searches,

Project Phoenix, has been up and running in its present form since 1995. Funded by an annual budget of $3 million, it is a little more than halfway through a ten-year survey of 1,000 Sun-like stars within 70 parsecs of Earth. Of the three largest radio SETI searches, it is the only professional search in which astronomers actually are at the radio telescope's controls in real time during the observing process. At this writing, Project Phoenix is using the Arecibo radio telescope in Puerto Rico during two annual three-week-long sessions to look for drifting continuous wave (CW) signals at high-resolution, which could include leakage from an alien civilization's day-to-day communications. Using 28 million channels simultaneously, Project Phoenix's observational strengths lie in the effective marriage of the radio telescope's hardware with the project's own software design and technology. This gives the search a sensitivity that would allow it to detect a cell phone signal on Jupiter, while at the same time making use of the project's massive catalog of interference signals to discount repeated false alarms.

VERIFYING THE REAL DEAL

Since Drake began his search over 40 years ago, radio astronomy has been increasingly hindered by interference from signals produced by the burgeoning satellite and telecommunications industries. Consequently, SETI astronomers must perform the arduous task of separating the electronic wheat from the chaff by logging in to their vast database to rule out terrestrial interference. If, for example, the radio telescope at Arecibo detects something interesting, astronomers must first verify that the source of the signal is from around the star and not terrestrial interference. To make that verification, Arecibo is pointed first toward the star and then away from the star. If the telescope can acquire the signal at any point on the sky, the Project Phoenix astronomers know that it is local interference. If after several tries the signal disappears while off the star, and is then reacquired while on the star, it becomes a serious candidate for evidence of extraterrestrial intelligence (ETI).

The "real thing" would of course require verification, first using independent software and data processing. This stage of the process is carried out at the University of Manchester's 76-meter Jodrell Bank radio telescope in the U.K., simply because if the signal really were from an ETI, then Jodrell Bank would observe it as well. However, due to Earth's own rotation, the U.K. telescope would receive the signal at a slightly different frequency. If the signal passes this test, the next phase would require days, possibly weeks, of verification, involving some 100 astronomers at multiple observatories.

Other groups are also searching for proof of ETI. Project Serendip, run by the University of California at Berkeley, uses data from Arecibo too, but takes a totally different approach. Instead of operating in real time, it piggybacks on day-to-day observations at Arecibo, looking for signals from ETI in data generated by radio astronomers making observations totally unrelated to ETI searches. Project Serendip's task is incredibly difficult. During a recent four-year search, the project was able to analyze data taken from observations of 93 percent of the observable sky. No ETI signals were found. But even if Serendip found a strong candidate signal, because its data analysis is not conducted in real time and sometimes takes place months after the observations, it would be unlikely that any found signal could be reacquired—thus further complicating the already difficult task of verification.

Nevertheless, Project Serendip remains undeterred and continues its work. Although it has not scored an ETI find, it has scored a big public relations coup among PC users. In the spring of 1999, it began its SETI@Home initiative, offering free software that could be downloaded as a PC screensaver. This unique program contains chunks of unanalyzed data from Serendip, which are then "crunched" when the owner's computer is not in use. PC owners periodically upload the crunched data back to an Internet site, where it is combined and recomputed on Serendip's own computers. An estimated 2,000 new data crunchers are signed on each day, and the ongoing project now has more than 2 million participants in more than 200 countries. The Serendip project also has a sibling "down under," called Project Southern Serendip, run by the University of Western Sydney Macarthur, which will be using Australia's 64-meter Parkes radio telescope in much the same manner as its counterpart in the Northern Hemisphere. SETI@Home will soon be expanded to include data from Project Southern Serendip.

"I view the SETI people as trying for a home run," says Geoffrey Marcy, "whereas a lot of us make a living by hitting singles and doubles. But if the SETI people are able to hit the home run and do it, then they are going to blow the rest of us away." For the time being, SETI remains in contention, and its advocates are eager to expand the playing field. Both Projects Phoenix and Serendip have targeted some of the newly discovered extra-solar planets and are searching for Earth-like planets that might support communicating technological civilizations. While thus far there's been only interference or silence, SETI radio astronomers remain undaunted. "We haven't even begun to look," says Jill Tarter, director of SETI research at the SETI Institute in Mountain View, California. "We're a very young technology in a very old galaxy, and this may take a while."[5]

Reaching Farther

For all its strengths, the Arecibo radio telescope is not ideal for SETI. It can scan only a few hundred star systems per year and has limited ability to steer; therefore, the radio SETI community is counting on new telescopes to help improve the odds of finding extraterrestrial intelligence. To that end, the University of California at Berkeley and the SETI Institute are collaborating on the $26-million One Hectare Telescope (1HT) that would have a total collecting area of 10,000 square meters (or 2.47 acres). 1HT would be capable of doing conventional radio astronomy and SETI searches simultaneously. It is due to go into full operation by 2005 at Berkeley's Hat Creek Observatory, a "radio-quiet" area near Mount Lassen in the California Sierra Madre mountains. Upon completion, the new telescope will be officially named the Allen Telescope Array, in tribute to Microsoft co-founder Paul Allen who is giving $11.5 million to the project. A seven-dish prototype is already in operation at the Russell Reservation, a wild-land

An artist's conception of the Allen Telescope Array, due to go into full operation by 2005 in the California Sierra mountains northeast of San Francisco. With its eventual collection of more than 1,000 antennas, its SETI target list will ultimately number as many as a million stars. (Ly Ly/SETI Institute.)

ecological research station a few miles from the university. The telescope array, initially consisting of 500 antennas of 5 meters each, will be able to scan several SETI targets for continuous wave signals at once, and then be expanded to include more than 1,000 dishes. Targets on the array's list will eventually number up to a million stars.

An even more ambitious international effort is behind the Square Kilometer Array (SKA), a coordinated group of radio telescopes with a collecting area of 1 million square meters, composed of 1,000 interferometrically linked antennas. Like the 1HT, the SKA would be used for both SETI and conventional radio astronomy. Since 1997, teams from Europe, the U.S., Canada, China, India, and Australia have been drawing up separate plans for this estimated $600-million project, which is slated to begin construction in 2010 and is scheduled for completion by 2015. A likely design would involve 30 interferometrically linked 200-meter diameter parabola-shaped radio telescopes, enabling SETI astronomers to detect what SKA's steering committee terms a "modest radio beacon from a significant fraction of our galaxy."

Sites under consideration for this ambitious project include the Upper Gascoyne-Murchison region of Western Australia, southern New Mexico, and the karst region of southwest China, an area noted for its natural limestone depressions. As Jill Tarter points out, the final decision will hinge on cost: "It's all about who can come up with a cost-effective way of building this thing for only several hundred dollars per square meter. Geopolitically, Australia is probably the only place in the world where we could still legislate a radio-quiet zone and eliminate transmitters on the ground. Having done that on the ground, we [will] have to deal with satellites. That's one of the reasons we think that the SKA should be made with 1,000 rather than 30 stations, because having those multiple stations [will] help to discriminate against interference."

Running Interference

In spite of these far-reaching plans, radio SETI and radio astronomy in general are facing a problem, which, if left unchecked, could make much of their efforts ineffectual: some radio astronomers predict that by 2010, terrestrial signal interference will reach a critical mass. This will effectively force radio astronomers to move into space to continue their observations. Recognizing this, even more far-flung sites for radio SETI have been proposed. In the early 1990s, the late Jean Heidmann, an astronomer at Observatoire de Paris-Meudon, was the first to suggest that the Moon's 80-kilometer wide Saha crater be used as a staging area for a SETI radio telescope. From its vantage point on the lunar far side, the Saha crater would provide proponents of conventional radio astronomy and SETI searchers with the ultimate lunar shield from Earth's plethora of radio frequency interference. Located just south of the lunar equator, Saha is only 350 kilometers from the Mare Smythii, a lunar mare (or smooth area of lowland lava plains) in full view of Earth. The biggest problem with situating such a telescope on the lunar far side is how to upload and download instructions and data to and from the telescope. One solution would be to use Mare Smythii as a relay station for data being sent down to Earth. Launching a small communications satellite into lunar orbit is a more plausible alternative. Such a lunar satellite would receive uploaded data from the Saha radio telescope when passing around the lunar far side and then download it back to Earth during its near-side passes.

Longtime SETI advocate Claudio Maccone, an astrophysicist and space scientist at Alenia Aerospazio in Turin, Italy, believes the success of a SETI lunar mission depends on an international accord to keep the lunar far side RFI-free. Should privately financed Moon exploration become commonplace, Maccone fears, not even the Moon's far side will remain immune to interference. His fears are not unjustified: proposals advocating the positioning of communications relay-satellites beyond the Moon have already been made. These satellites would enable real-time Earth uplinking of data from vantage points some 60,000 kilometers over and beyond the Moon's far side. Maccone himself has proposed sending an unmanned $400-million space mission to deploy an inflatable radio telescope inside the Saha crater. The telescope would come attached to a 350-kilometer weighted, fiber-optic cable that would link the radio telescope with a radio relay antenna in the Mare Smythii, which would be capable of two-way data transfer back to Earth, all without disturbing the Saha crater's interference-free status.

The basic plan involves sending a spacecraft loaded with two separate payloads into a lunar equatorial orbit. Once in lunar orbit, the spacecraft would split into two separate craft, or payloads, each of which would make their way toward separate landing sites on the lunar far side. One payload would be loaded with the relay antenna, and the other with the radio telescope, possibly a 10-meter inflatable parabolic dish. A weighted tether would be deployed between the now separate craft at an altitude of 160 kilometers above the far side's equator. First, the two separate craft would stabilize in lunar orbit, and the tether between them would extend to a distance of 350 kilometers. Next, each craft would fire its guidance thrusters into a descent for its target, either the Saha crater or Mare Smythii. By this point, the Moon's gravity

NASA's Galileo spacecraft took this image of the Far Side of the Moon, which has been targeted as a prime spot for SETI due to its shielding from Earth's radio frequency interference. (NASA/JPL/Caltech.)

would have caused each payload's orbit to slowly decay, to the extent that the thrusters could be used for positioning them for landing. Maccone believes that, with a robust design, each of the payloads should survive impact on the lunar surface. Once the antenna and the telescope had been successfully deployed on the lunar surface, the fiber optic cable encased in the tether would link the far-side telescope with the Earth-targeted relay station in the Mare Smythii. By forgoing a manned mission, Maccone estimates that his plan could take wing within the next two decades. The technology is already available, as both the antenna and the telescope would opt for soft landings using airbags similar to NASA's Mars Pathfinder mission. The radio telescope would share its resources for regular radio astronomy, which Maccone hopes will be a selling point. A lunar site might be all that radio SETI needs. After all, aside from a noble effort and improved technology for radio astronomy as a whole, more than 40 years of Earth-bound searching for radio signals from extraterrestrial intelligence has resulted in little more than static—and a steady stream of overwrought media coverage.[6]

ALTERNATIVE APPROACHES

As University of Arizona astronomer Neville Woolf points out, looking for evidence of ETI in the radio spectrum could well be futile, because the ETI may have long moved on to more advanced forms of communication, which our own technologically primitive civilization has yet to realize. "Radio SETI is a noble search," says Woolf. "Yet, if we were to look for [ETI] technologies by looking for giant steam engines, people would laugh because they would say that's old technology. But what's old on the scale of a Universe that's 10 billion years old?"[7]

Even Frank Drake believes that it is time to broaden the search to other parts of the electromagnetic spectrum. In an effort to extend the search parameters, SETI astronomers are now looking in the optical spectrum for ETI signals that would appear in either the infrared or visible wavelengths in the form of narrowband laser emissions. Optical SETI (or the search for ETI in the optical spectrum) was first proposed in April 1961 by the Nobel-winning Berkeley physicist Charles Townes soon after he shared the patent for the design of the laser.

The power of the laser (light amplification by stimulated emission of radiation) lies in its capability to move photons of light in coherent monochromatic beams. The laser produces light of the same wavelength, while also maintaining coherency (allowing photons to propagate with phase constancy), so that the distance between the crests and troughs always remains constant. Townes realized that if we have the ability to utilize the physics of the laser, then perhaps ETI would also be using them and would prefer the optical spectrum over radio when communicating over interstellar distances. Few in the astronomical community took the idea of optical SETI seriously until it became clear that it would be possible to produce high-powered information-laden laser pulses capable of attracting the attention of nearby extraterrestrial civilizations.

Already, some astronomers argue that at optical frequencies, narrow pulse lasers can convey more information per received photon than radio waves. And it has been proven that a nanosecond pulse, from a laser produced by ETI and using only a modest telescope, would easily outshine the civilization's parent star. If so, then, in theory, on any given clear night, we may in fact be witness to intelligent optical signals that we now simply mistake as random starlight. But even if such extraterrestrial laser communications exist and are detectable, the challenge of actually spotting them remains a Herculean task.

The Advent of Optical SETI

The University of California at Berkeley has two ongoing optical SETI programs, one of which is overseen by Marcy and simply piggybacks on data taken from Marcy and Butler's 1,000-star Doppler spectroscopy survey. They hope to find sharp, continuous laser signals in the spectra of their extra-solar planet searches. At the same time, Dan Werthimer, another Berkeley astronomer, is leading a search for pulsed laser signals around 2,500 nearby F, G, K, and M stars, as well as around a few globular clusters and galaxies. In his effort, Werthimer uses Berkeley's 30-inch (76-centimeter) automated telescope at Leuschner Observatory. At the Harvard Smithsonian Center for Astrophysics, planet hunters David Latham and Robert Stefanik are, like Marcy and Butler, using their 2,500-star Doppler spectroscopy survey as an opportunity to do optical SETI—in their case, using a 1.5-meter telescope in Harvard, Massachusetts. The Harvard team is also planning an all-sky optical SETI effort. And halfway around the world, Ragbir Bhathal, an astronomer at the University of Western Sydney, is directing the Australian optical SETI project as part of a five-year search of southern Sun-like circumpolar stars, southern globular clusters, and a few galaxies. Located at the Campbelltown Rotary Observatory, they are looking for nanosecond ETI pulses using two telescopes of 30 and 40 centimeters.

Facing Reality

In truth, no matter how dogged our efforts to find extraterrestrial intelligence, and how advanced the technologies, SETI astronomers may still miss the mark. "Back-of-the-envelope" calculations on the median age of possible extraterrestrial civilizations suggest their technologies would be several million years ahead of ours. Thus, trying to detect their communications or signals in the radio and optical spectrums may indeed be as futile as trying to log on to the Internet by banging on a "talking" drum. How can we second-guess an extraterrestrial civilization that could be millions of years ahead of us, if we cannot accurately envisage what Earth technology will bring in only several hundred years? We've moved from basic laptop computers to wireless Dick Tracy-style communicating watches in 10 years. What will be wrought in 10 million years? As Stuart Bowyer, a retired Berkeley radio astronomer and longtime SETI advocate, reminded me, we are simply stuck with the physics that we understand thus far. Bowyer believes that even if other civilizations are very advanced, communication in both the radio and optical spectrums will remain viable, and, therefore, should remain central to SETI's overall approach. He asserts that in the next 40 years, or by the time NASA and ESA have produced high-resolution images of an extra-solar "earth," SETI astronomers will have had a "fair chance" of finding extraterrestrial technology.[8]

But even if a signal, or "leakage," is detected, it would be doubtful that we could decipher it, unless it was intended as an all-points-directed beacon designed to be decipherable for any civilization that lay in its beam. Certainly, signal detection would confirm that we were not alone in the Universe, but true interstellar communication, even over short distances, would necessitate a transgenerational attitude of collective long-term effort. Signals sent and returned over distances as short as 10 parsecs would require a round-trip communication transit of at least 65 years, or basically a human lifetime. Sustaining such interest in the public at large may prove to be optimistic at best. Imagine a student in a third-grade class some time in the future: the teacher announces that SETI astronomers have just detected a signal containing a message from an intelligent technology-bearing extraterrestrial civilization on an Earth-like planet circling a nearby star. The announcement would initially be greeted with emotions ranging from excitement, consternation, curiosity, and bewilderment. Universities would likely offer whole new curricula based on how Earth's religions and philosophies would be affected by the news. The Nobel Committee would award prizes, and the media would have a field day. But by the time the message had been deciphered, and the international community had agreed upon and sent humanity's response, the student would hear the news of ET's reply as he was "getting the gold watch" at his retirement party. So, dreams of interstellar E-mail are slightly optimistic. While profound certainly, the detection of an alien intelligence, even within 100 parsecs from Earth, would not be something that most people would think about on a day-to-day basis. Life would simply go on.

1   Gould, Stephen Jay 1989. *Wonderful Life: The Burgess Shale and the Nature of History*. New York: W. W. Norton: 48.

de Duve, Christian René 1995. *Vital Dust: Life as a Cosmic Imperative*. New York: Basic Books: 293.

2   Marcy, Geoffrey, astronomer, University of California at Berkeley. Interviewed on May 25, 1999, at Dana Point, California, and on August 6, 1999, at Hapuna Beach, Hawaii. Follow-ups took place on September 8, 2000, May 10–12, 2001, and June 3, 2001.

3   Lineweaver, Charles H. "Why Did We Evolve So Late?" Poster paper presented at Bioastronomy 99, Hawaii, August 6, 1999.

Livio, Mario. "How Rare Are Extraterrestrial Civilizations, and When Did They Emerge?" *The Astrophysical Journal* 511 (January 20, 1999): 429.

4   Cocconi, Giuseppe and Philip Morrison. "Searching for Interstellar Communications." *Nature* 184 (September 19, 1959): 844.

5   Tarter, Jill C., director of SETI Research, SETI Institute. Interviewed on August 5, 1999, at Bioastronomy 99, Hawaii.

6   Maccone, Claudio 1999. "Space Missions Enabling SETI Searches Farther and Farther Out." http://www.nidisci.org.essaycomp/cmaccone.

7   Woolf, Neville, astronomer at University of Arizona, Tucson. Interviewed on August 5, 1999, at Bioastronomy 99, Hawaii.

8   Bowyer, Stuart, astronomer, University of California at Berkeley. Interviewed on August 7, 1999, at Bioastronomy 99, Hawaii.

EPILOGUE

Halfway through the year 2001, NASA announced a new mission to the Solar System's inner-most planet Mercury, for launch in early 2004, and renewed concept studies for a possible mission to Pluto within the same time frame. The novel aspects of both missions remind us that even as the number and variety of planets known to circle other stars continue to mount, we have much to learn about our own Solar System. At the same time, we can be encouraged that astronomers continue to find remnants and parts of other planetary systems that appear to mirror the most salient parts of our own. In February 2001, infrared observations made at the Keck Observatory by UCLA astronomers Christine Chen and Michael Jura indicated that a massive asteroid belt orbits the young (estimated 100 million-year old) star Zeta Leporis, some 21 parsecs away from Earth in the constellation Lepus.[1] Then, in June 2001, the Space Telescope Science Institute (STScI) reported that Hubble observations of microlensing events in the Globular Cluster M22, some 2,600 parsecs away in Sagittarius, picked up the signatures of free-floating planetary-type objects with minimum masses only 80 times that of Earth. As reported in *Nature*, monitoring 83,000 background stars behind M22 between February and June 1999, the team observed six events in which a background star jumped in brightness by some 50 percent for less than 20 hours. This short brightening of the background star indicates the foreground microlens is being created by a very low-mass object, not unlike an ejected planet. If these observations can be verified, these objects are very likely planets that have been "detached" (or gravitationally ejected or torn) from their orbits around their parent stars. To confirm these events, the team is planning more observations over a continuous seven-day period.[2]

REAL JUPITERS?

Yet the most tantalizing news of all may come by the end of 2001. At this writing, Geoffrey Marcy's team is tracking two planets each with a minimum of 1 Mj at distances approaching 5 AU from their parent stars. One of these planetary candidates circles 55 Cancri, a G8 star 12.5 parsecs away from Earth in the Canceris constellation. 55 Cancri is already known to have a giant inner gas companion circling its parent star once every 14.6 days, as was announced by Marcy and Butler in 1996. However, as Marcy pointed out to me in early May 2001, his team's observations initially indicated with a 97-percent confidence level that there was an outer planet circling 55 Cancri with an 11-year orbital period, equal to a distance of 4.8 AU from its parent star. But until they have obtained a 99-percent confidence level, the team will not publicly announce that the two stars in question do indeed have planets with Jupiter-like periods. If the outer planet around 55 Cancri should eventually be confirmed by the astronomers, it would lie at the farthest distance from its star of any previously discovered extra-solar planet.[3]

As stressed earlier, if we find real "Jupiters" with highly eccentric or elliptical orbits, it might not bode well for the creation of Earth-like planets with stable orbits. As for these two "real" Jupiter candidates, Marcy would only say: "We don't quite know yet whether the 'Jupiters' at 5 AU that we are about to secure have circular or eccentric orbits. The orbits aren't well enough defined yet. It will be another few months or longer before we know."

Even without the confirmation of a Jupiter-mass planet at 5 AU from 55 Cancri, it has been argued that this star could harbor the nearest analog to our own Solar System of any star thus far surveyed. Estimated to be at least 3 billion years old, 55 Cancri was first reported to emit an excess of far infrared radiation in 1998 in surveys by ESA's Infrared Space Observatory. Then, that same year, two University of Arizona astronomers, David Trilling and Robert H. Brown, reported that observations using a coronagraph on NASA's Infrared Telescope Facility atop Mauna Kea, Hawaii, had allowed them to image a disk around 55 Cancri. According to their observations, this disk extends from 27 to 45 AU and is suggestive of a mature planetary system, although the inner and outer edges are not known precisely, as the coronagraph blocked their view of the disk's inner edge. But the team reported that the disk had a spectral signature similar to that of our Solar System's Kuiper Belt, indicating that it must contain methane ice, as found in Pluto and the Kuiper Belt. Thus, Trilling and Brown suspect that 55

Cancri's disk must be full of the kind of leftover debris expected to be found at the outer edge of a fully mature planetary system.[4]

In some ways, the last decade of planet-hunting has been like the early days of pulsar studies nearly 35 years ago. It took time to understand the true nature of what the observations were revealing. And until a real consensus on any given astrophysical phenomenon can be reached and become ingrained within the astronomical community's collective psyche, there will dissenting points of view. Thus, in late 2000, David Black of the Lunar and Planetary Institute in Houston, along with George Gatewood of the Allegheny Observatory, and Inwoo Han of the Korea Astronomy Observatory, presented preliminary data on an astrometric study of 30 stars observed by the Hipparcos spacecraft. All 30 of the stars have also been identified by Doppler spectroscopy surveyors as harboring planets. Their findings were presented at the Division for Planetary Sciences of the American Astronomical Society meeting in Pasadena, where again Black contended that the orbital periods and eccentricities of many of these so-called extra-solar planets are indistinguishable from binary stars. Black believes that the Doppler spectroscopy searches have orbital inclinations that are extremely low, which according to Black means that "the orbital planes of these companions appear to be oriented nearly face on to the observer." If so, the planetary companions' actual masses would not be 30 percent more than the minimum masses ascribed by the Doppler spectroscopy teams, but instead would be many times larger.[5]

For instance, in the case of 55 Cancri, Marcy and Butler first reported in 1996 that the inner companion had a minimum mass of 0.84 times that of Jupiter. But data on the orbital inclination of the 55 Cancri circumstellar disk imaged by Trilling and Brown suggest that the mass may, in fact, be 1.9 times that of Jupiter. Yet the Black/Gatewood/Han study indicates that this inner companion's true mass may be higher still, perhaps making it a brown dwarf. So, if many of these Doppler spectroscopy planets are in reality stellar or brown dwarf companions of binary or multiple star systems, how would they come to have such short period orbits around each other? As Gatewood points out, some binary stars are so close that they are in actual physical contact, making their orbital periods extremely short.[6] Floor van Leeuwen at the Cambridge University notes that such short period binaries are mainly formed within star clusters. Their short periods are frequently due to interactions between a binary and a third star. Van Leeuwen says that the third star is often ejected from the cluster, while two of the stars will remain in very close orbit around each other.[7]

"Some of the [extra-solar planetary] candidates are in fact binary stars," says Gatewood. "Some of the remaining candidates will be the result of an effect, perhaps associated with stellar age or metallicity, that is not yet understood. This effect may produce the orbiting bodies that we think we have detected, or it may simulate the radial velocity motions we are seeing. I note that many of the motions do not seem to be easily modeled by a single companion. If you propose two or more companions you allow yourself so many degrees of statistical freedom that you can model a wide variety of various types of motions. Some of these may actually be associated with the star's atmosphere and not with an orbiting companion. And some of the objects will be actual planets."

Gatewood believes that in fact only half of the 30 Doppler spectroscopy candidates that they included in their astrometric study may indeed be *bona fide* planets. Yet even those with such contrary points of view believe the galaxy and, indeed, the Universe is awash with planets—planets of all sizes, shapes, orbits, around all kinds of stars, as well as Earth-like planets in habitable orbits. But again, is there a quicker way to determine their real number?

THE GOLDILOCKS EFFECT

In the Spring of 2001, Charles Lineweaver, a physicist at the University of New South Wales in Sydney, put forth a theoretical model that uses what is known as a Goldilocks selection effect to estimate the number and probable ages of Earth-like planets in the Universe. As applied to planet formation, the Goldilocks effect is a variation on the well-known fairy tale's notion that the porridge is either "too hot," "too cold," or "just right." Lineweaver theorizes that given the gradual buildup of metals in the Universe, three quarters of the Earth-like planets in the Universe are older than our Earth. (Remember from Chapter 1 that astronomers term "metal" any element with an atomic weight heavier than helium.) Through spectroscopy, astronomers are able to measure a star's metallic makeup, thus enabling them to determine the types and proportions of metals. Lineweaver has attempted to take known stellar metallicities and formulate a coherent correlative theory of how stellar metallicity affects the formation of Earth-like planets not only in our local neighborhood, but throughout the Universe. Based on the estimated peaks in the Universe's distribution of stellar metallicities, he puts the average age of extra-solar "earths" at 1.8 billion years older than our own Earth. This figure is derived, in part, on the following axioms:

- A star with zero metallicity (or no metals) would lack the metals to form an Earth-like planet;

- A star with too high a metallicity would have a highly metal-rich protoplanetary disk and thus lend itself to making an overabundance of gaseous giant planets that would gravitationally impede or prohibit the formation of "earths";

- A star with metallicities somewhere between these two extremes would be "just right" and tend to form Earth-like planets in stable orbits. Our Sun is estimated to have more metals than two-thirds of nearby Sun-like stars and less metals than two-thirds of stars thought to host so-called hot Jupiters. Thus, the Sun may have a natural Goldilocks effect.[8]

Van Leeuwen, by contrast, acknowledges that metallicity is "probably a factor in the whole [planet formation] process," but believes there are more important factors in determining whether a star is a likely candidate to harbor Earth-like planets. "The problem with a lot of these theories of [planetary] system formation," says van Leeuwen, "is that there is always a link missing, which has to be put in more or less artificially."

Still, Lineweaver maintains that 10 percent of stars in the "entire Universe" would harbor habitable "earths." However, at the moment, Lineweaver says he is looking into just how many nearby stars are likely to harbor Earth-like planets. Unlike our G2 star, most of the stars within 10 parsecs of Earth are of the M spectral type, meaning that they are smaller and less luminous, thus making their habitable zones smaller.

But full-fledged planetary systems harboring habitable "earths" must exist. The processes of planet formation around other stars are too much in evidence for this not to be so. In fact, in some small way, the ultimate fate of our species might depend upon these most recent, rudimentary series of informal "planetary censuses."

STELLAR ENDGAME

The long evolutionary journey from paramecia to astronaut may be as inextricably linked to the survival of intelligent life in the galaxy as our own desire for self-preservation. (Our distant progeny living on a nearby Earth-like planet may eventually figure it out.) But even though we don't know where our descendants will be 500 million years from now, scientists do have a good idea of what will be happening to the Sun.

We know that as the Sun carries on its fusion of four hydrogen atoms into one atom of helium, the Sun's helium core also gets hotter and hotter, causing the Sun to grow in luminosity and radiation by an average of 10 percent every 1 billion years. Eventually, the Sun's region of hydrogen fusion will encroach on the solar photosphere (its visible surface), and once the burning is within proximity of the solar surface, the Sun will have lost 40 percent of its mass. But long before that happens, the damage will have been done here on Earth. To continue the cycle of life as we know it 500 million years hence, humankind will be forced to leave our planet to avoid the Armageddon of our own G2 star. Within 500 million years, the heat of the Sun will cause carbon dioxide to fall below its compensation point, the limit necessary for plants to conduct photosynthesis. The result? Death to all of Earth's biota. Beyond that, the oceans are expected to evaporate into the vacuum of space between 900 million and 1.1 billion years from now, and 5 to 6 billion years later, the Sun will enter its red giant phase. That's when its supply of hydrogen will have been exhausted and it will expand to a diameter 25 times its present size. This phase will last for an estimated half a billion years. Not even icy Pluto will escape the Sun's withering heat, but will finish its life like a snow cone melting on a hot summer's day. Earth will inevitably become entangled in the endgame mechanics of our dying Sun. When that happens, the Sun will either swallow Earth whole or give it the bum's rush from its present orbit to the edges of the Solar System. In the bitter end, the outer half of our star will be dispersed back into the interstellar medium, while its core will collapse into a dense low-luminosity core, or

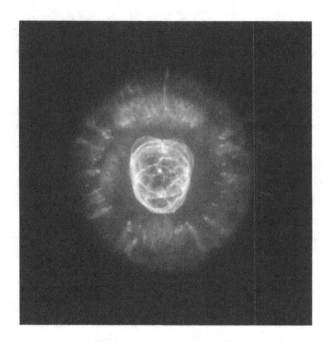

Ejected remnants from our Sun's red giant phase will form planetary nebulae, closely resembling the Eskimo Nebula (NGC 2392). First discovered by William Herschel in 1787, this 10,000-year-old nebula lies an estimated 1,000 parsecs away from Earth in the constellation Gemini. This Hubble Space Telescope image clearly shows where the Eskimo Nebula got its name: some think it resembles a face surrounded by a fur parka. (NASA, A. Fruchter, and the ERO Team [STScI].)

white dwarf, approximately the size of Earth. Ejected remnants from the Sun's red giant phase may, in fact, become fodder for yet a new planetary system. Astrophysicists mining data collected from ESA's ISO satellite have found 100,000 heretofore-unseen red giant stars in a region near our own galaxy's center, indicating that our Sun's evolution is certainly not uncommon.

EXTRA-SOLAR ENDINGS

As a reminder of our Solar System's ultimate fate, NASA's Submillimeter Wave Astronomy Satellite (SWAS), launched in 1998, has identified spectral lines of H_2O indicating that massive amounts of water ice are being vaporized around CW Leonis, a giant M star more than 150 parsecs away from Earth in Leo. As the star undergoes its final death rattle in its post red giant phase, the SWAS science team reported in July 2001 that the expanding star has vaporized and is continuing to vaporize water ice from the surface of several hundred billion comets. This process has created massive vapor clouds up to 100 AU across at distances of up to 300 AU from the star.[9] A billion years ago, CW Leonis very likely had billions of comets in orbit around what was a stable Sun-like main sequence star, only 1.5 to 4 times as massive as the Sun. However, as David Neufeld, SWAS team member and an astrophysicist at The Johns Hopkins University, points out, "We have no idea whether there were any planets orbiting this star. But before this, there was only one system [known] to have ice, and that is our Solar System."[10]

GETTING THERE

While we may be on the verge of detecting habitable planets, planets that may harbor life, or planets that ultimately may be candidates for colonization, we are most likely a few centuries away from launching a spacecraft that could get us there in the flesh. To date, after circling around our own Solar System for the last few decades, we've barely met the boundaries of our own "backyard" and can only wonder how to leave the neighborhood for good.

Pioneer 10, a 260-kilogram interplanetary NASA probe launched to Jupiter in 1972, is the first human-made object to venture beyond the orbit of Pluto. Now more than 7 billion miles from Earth, it is traveling at 13 kilometers per second and still deriving its energy from decaying plutonium 238—enough to generate an 8-watt signal, about that of a child's night-light. On its present course, some 30,000 years from now, the probe will pass within three light years of Ross 248, an M star 3 parsecs away from Earth in Taurus, a winter constellation visible in the Northern Hemisphere's southwestern sky just after sunset. However, NASA's Voyager 1 spacecraft, launched five years after Pioneer 10 for flybys of Jupiter and Saturn, is now classified by NASA as the "farthest human-made object from Earth."[11] It has traveled about twice the distance that Pluto lies from our own Sun. But it will likely be 2003 before Voyager 1 reaches the heliopause, a boundary region thought to lie some 100 AU from Earth, where the solar wind is no longer prevalent and interstellar space truly begins.

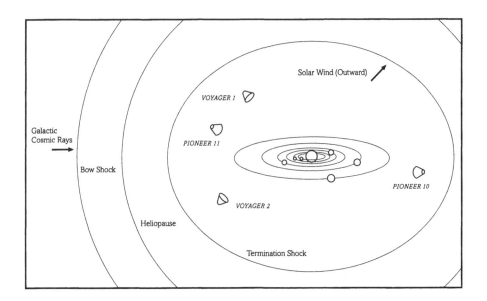

Several NASA spacecraft are reaching the boundaries between the outer reaches of the Sun's sphere of influence and interstellar space. Pioneer 10 is now making its way through the outer limits of the heliosphere, a bubble carved out of the gaseous interstellar medium by the solar wind. The heliosphere is at least twice as large as the orbit of Pluto, but its exact outer boundaries are still unknown. (Based on a graphic by NASA/JPL.)

Even given the restrictions of present-day physics, in 500 years time, it seems likely that humans will have found a way to get to the nearest stars and will be colonizing an Earth-like planet circling some nearby Sun-like star. Five hundred million years ago, trilobites were scavenging the seafloor of what is now southern Arizona. Their hardy exoskeletons were the ancient precursors of our skull and bones, without which we would never have evolved to walk upright. But we too are probably only an evolutionary "pit stop" *en route* to a higher life form.[12] In another five hundred million years, our own evolutionary descendants may be trading human fossils at some outpost of colonization light years away. It might even be in a planetary system that was first spotted in our current era by one of the students of Marcy or Mayor.

1 "Astronomers Identify Evidence of Asteroid Belt Around Nearby Star: Findings Indicate Potential for Planet or Asteroid Formation." UCLA press release (June 4, 2001).

2 "M22." Students for the Exploration and Development of Space (SEDS) chapter at the University of Arizona's Lunar and Planetary Laboratory. http://www.seds.lpl.arizona.edu/messier (December 9, 1999).

"Wandering Mystery Planets." *NASA Science News* (June 29, 2001).

"Hints of Planet-Sized Objects Bewilder Hubble Scientists." JPL press release (June 27, 2001).

Sahu, Kailash C. et al."Gravitational Microlensing by Low-Mass Objects in the Globular Cluster M22." *Nature* 411 (June 28, 2001): 1022.

3 Marcy, Geoffrey, astronomer, University of California at Berkeley. Private communication with the author on May 10–12 and June 3, 2001.

4 Trilling, D. E. and R. H. Brown. "Imaging a Circumstellar Disk Around a Star with a Radial-Velocity Planetary Companion." Paper presented at the AAS DPS meeting, August 11, 1998.

5 Alexander, Amir. "Extra Solar Planets or Small Stars?" *The Planetary Society Daily Online News*, http://www.planetary.org/html.news (October 31, 2000).

Han, Inwoo, David C. Black, and George D. Gatewood. "Preliminary Astrometric Masses for Proposed Extra-Solar Planetary Companions." Paper presented at the 2000 Meeting of the Division for Planetary Sciences of the American Astronomical Society, Pasadena, California, October 23–27, 2000.

"Some New Planets Might Be Stars, Researchers Say." University of Pittsburgh news release (October 28, 2000).

6   Gatewood, George D., astrometrist, Director of Allegheny Observatory, University of Pittsburgh, Pennsylvania. Private communication with the author on June 25, 2001.

7   van Leeuwen, Floor, astronomer, University of Cambridge, U.K. Interviewed on October 29, 1999, and on June 26, 2001.

8   Lineweaver, Charles H. "An Estimate of the Age Distribution of Terrestrial Planets in the Universe: Quantifying Metallicity as a Selection Effect." *Icarus* 151 (2001): 307.

      Lineweaver, Charles H., physicist, University of New South Wales, Sydney, Australia. Private communication with the author on June 28–29, 2001.

9   "NASA Establishes Contact with Famed Pioneer 10 Spacecraft." NASA Ames Research Center, Moffett Field, California, press release (April 30, 2001).

      "Most Distant Spacecraft May Reach Shock Zone Soon." NASA JPL press release (December 18, 2000).

10   Neufeld, David, SWAS team member and astrophysicist, The Johns Hopkins University, Baltimore, Maryland. Interviewed on July 12, 2001.

11   "Most Distant Spacecraft May Reach Shock Zone Soon." NASA JPL press release (December 18, 2000).

      "NASA Establishes Contact with Famed Pioneer 10 Spacecraft." NASA Ames Research Center, Moffett Field, California, press release (April 30, 2001).

12   de Duve, Christian René 1995. *Vital Dust: Life as a Cosmic Imperative*. New York: Basic Books: 301.

"A Puzzle of Galactic Evolution Is Solved: Massive Gas Clouds Seed the Galaxy with the Stuff of Stars." Space Telescope Science Institute press release (November 24, 1999).

"A Viking Galaxy Found by Hipparcos Shows How the Milky Way Grew," *ESA Science News*, http://sci.esa.int (November 5, 1999).

Adler, Robert. "Beyond the Stars." *New Scientist* 164, no. 2217 (December 18, 1999): 4.

Alexander, Amir. "Extra Solar Planets or Small Stars?" *The Planetary Society Daily Online News*, http://www.planetary.org/html.news (October 31, 2000).

"Ancient Meteorite Hints at Sol's Birth." *Space Daily Express*, http://www.spacedaily.com (May 29, 2000).

"Ancient South African Soils Point to Early Terrestrial Life." Penn State University press release (November 29, 2000).

"Andromeda Galaxy: An Infrared View of Globulars in the Andromeda Galaxy." *Astronomy Magazine Online*, http://www.astronomy.com (January 28, 2000).

Angel, Roger 1998. "Search and Analysis for Substellar and Planetary Companions by Direct Detection." In *Failed Stars and Super Planets: A Report Based on the January 1998 Workshop on Substellar-Mass Objects*. Space Studies Board, National Research Council. Washington, D.C.: National Academy Press: 15.

Annis, James. "An Astrophysical Explanation for the Great Silence." *Journal of the British Interplanetary Society* 52 (January 1999): 19.

Announcement of Upsilon Andromedae Multiple System press conference at San Francisco State University (April 15, 1999). Author's notes.

"Are Solar Systems Like Ours Rare?" *Spaceviews.com* E-mail newsletter (January 30, 2000).

Armitage, Philip J. and Brad M. S. Hansen. "Early Planet Formation as a Trigger for Further Planet Formation." *Nature* 402 (December 9, 1999): 633.

Armstrong, Thomas. "Stellar Optical Interferometry in the 1990s." *Physics Today* 48 (May 1995): 42.

"Astrobiologists Find Evidence of Early Life on Land." NASA Ames Research Center press release (November 29, 2000).

"Astrobiology in the U.K." Community Report to the British National Space Centre (October 1999).

"Astronomers Discover Six New Planets Orbiting Nearby Stars." University of California at Santa Cruz press release (November 29, 1999).

"Astronomers Find Evidence for the First Planet Seen Orbiting a Pair of Stars." National Science Foundation press release (November 3, 1999).

Backer, D. C. et al. "A Second Companion of the Millisecond Pulsar 1620 –26." *Nature* 365 (October 1993): 817.

Bailey, Jeremy. "Circular Polarization in Star-Formation Regions: Implications for Biomolecular Homochirality." *Science* 281 (July 31, 1998): 672.

Baliunas, Sallie L. et al. "Properties of Sun-like Stars with Planets: Rho Cancri, Tau Boötis, and Upsilon Andromedae." *The Astrophysical Journal* 474 (January 10, 1997): L119.

"Barnard's Star." Department of Physics, University of Durham, U.K., http://www.dur.ac.uk/~dph0jrl/one_lab/pm_barn.html.

Barrow, John D. and Frank J. Tipler 1986. *The Anthropic Cosmological Principle.* Oxford, U.K. & New York: Oxford University Press.

Basri, Gibor 1998. "The Physical and Orbital Characteristics of Known Substellar-Mass Objects." In *Failed Stars and Super Planets: A Report Based on the January 1998 Workshop on Substellar-Mass Objects.* Space Studies Board National Research Council. Washington, D.C.: National Academy Press: 21.

Beckwith, Steven V. W. and Anneila I. Sargent. "Circumstellar Disks and the Search for Neighboring Planetary Systems." *Nature* 383 (September 12, 1996): 139.

Beichman, C. A., Neville J. Woolf, and C. A. Lindensmith. "The Terrestrial Planet Finder (TPF): A NASA Origins Program to Search for Habitable Planets." The TPF Science Working Group, TPF/JPL publication 99-3 (May 1999).

Bell, George H. "The Search for the Extra-Solar Planets: A Brief History of the Search, the Findings and the Future Implications." http://www.public.asu.edu/nsciref/explont.htmll. Posted in May 1997, revised December 19, 1997.

Bell Burnell, Jocelyn. "After-Dinner Speech at the Eighth Texas Symposium on Relativistic Astrophysics." *Annals of the New York Academy of Science* 302 (1977): 685.

Bennett, D. P. et al. "Gravitational Microlensing Evidence for a Planet Orbiting a Binary Star System." *Nature* 402 (November 4, 1999): 57.

Bhathal, Ragbir. "Australian Optical SETI Project." Paper presented at Bioastronomy 99, Hawaii, August 6, 1999.

Bhathal, Ragbir. "The Case for Optical SETI." *Astronomy & Geophysics* 41, no. 1 (February 2000): 25.

Black, David C. and Tomasz Stepinski. "A Statistical Assessment of the Assertion That Low-Mass Companions to Stars Are Extra-Solar Planets." Abstract presented at the From Giant Planets to Cool Stars conference, Flagstaff, Arizona, June 8–11, 1999.

Blum, Jürgen et al. "Growth and Form of Planetary Seedlings: Results from a Microgravity Aggregation Experiment." *Physical Review Letters* 95, no. 12 (September 18, 2000): 2426.

Borucki, W. J. et al. "The Vulcan Photometer: A Dedicated Photometer for Extra-Solar Planet Searches." Abstract from the American Astronomical Society Annual Meeting, Atlanta, Georgia, January 2000.

Boss, Alan P. "Giant Giants or Dwarf Dwarfs." *Nature* 409, no. 6819 (January 25, 2001): 462.

Boss, Alan P. "Giant Planet Formation by Gravitational Instability." *Science* 276 (June 20, 1997): 1836.

Boss, Alan P. "The Birth of Binary Stars." *Sky & Telescope* 97 (June 1999): 32.

Boss, Alan P. "Twin Planetary Systems in Embryo." *Nature* 395 (September 24, 1998): 320.

Boston, William. "New Frontiers: Putting 'Cyber' into Space Could Open Up a New World." *Wall Street Journal Europe/Convergence* supplement (Winter 1999): 15.

Bouchy, François and P. Connes. "Autoguider Locked on a Fiber Input for Precision Stellar Radial Velocities." *Astronomy & Astrophysics* Supplement 136 (April 1, 1999): 193.

Brack, André and Gero Kurat. "Artificial Meteorites: Foton-12 Results." *On Station*, issue 2 (March 2000): 17.

Bremer, Michel, radioastronomer, Institute for Millimetric Radioastronomy, Grenoble, France. Interviewed on July 13, 1999, at La Silla, Chile.

Bridges, Andrew. "Interferometry Mission Sent Back to Drawing Board." *Space.com* via *Spaceviews*, http://www.spaceviews/2000/11/13b.html (November 13, 2000).

Britt, Robert Roy. "Study: Solar Systems Common." http://www.space.com/scienceastronomy/astronomy/gas_giants (January 4, 2001).

Brown, Timothy M. et al. "A Search for Line Shape and Depth Variations in 51 Pegasi and Tau Boötis." *The Astrophysical Journal* 494 (February 10, 1998): L85.

Burrows, Adam 1999. "Extra-Solar Giant Planet and Brown Dwarf Theory." In *Planets Outside the Solar System: Theory and Observations*. J. M. Mariotti and D. Alloin (eds.). Dordrecht: Kluwer Academic Publishers: 121.

Burrows, Adam and Roger Angel. "Direct Detection at Last." *Nature* 402 (December 16, 1999): 732.

Butler Burton, W. "Clouds from Near and Far." *Nature* 402 (November 25, 1999): 359.

Butler, R. Paul et al. "AAO Planet Search Program." Anglo-Australian Telescope home page, http://www.aao.gov.au.

Butler, R. Paul et al. "Evidence for Multiple Companions to Upsilon Andromedae." *The Astrophysical Journal* 526, no.2 (December 1999): 916.

Butler, R. Paul et al. "Three New 51 Pegasi-Type Planets." *The Astrophysical Journal* 474 (January 10, 1997): L115.

Charbonneau, David et al. "An Upper Limit on the Reflected Light from the Planet Orbiting the Star Tau Boötis." *The Astrophysical Journal Letters* 522 (September 10, 1999): L145.

Chown, Marcus. "The Last Supper." *New Scientist* 164, no. 2212 (November 13, 1999): 44.

Cocconi, Giuseppe and Philip Morrison. "Searching for Interstellar Communications." *Nature* 184 (September 19, 1959): 844.

Cohn, David V. 2000. "The Life and Times of Louis Pasteur." http://www.labexplorer/louis_pasteur.htm. Adapted from a 1996 keynote address at the University of Louisville, Kentucky, marking the centennial celebration of Pasteur's death.

Colavita, M. M. 1999. "Astrometric Techniques." In *Planets Outside the Solar System: Theory and Observations*. J. M. Mariotti and D. Alloin (eds.). Dordrecht: Kluwer Academic Publishers: 177.

Collier Cameron, Andrew et al. "Probable Detection of Starlight Reflected from the Giant Planet Orbiting Tau Boötis." *Nature* 402 (December 16, 1999): 751.

"Conan the Bacterium." *Space Daily Express*, http://www.spacedaily.com (December 14, 1999).

Connes, P. et al. 1996. "Demonstration of Photon-Noise Limit in Stellar Radial Velocities." *Astrophysics and Space Science* 241. Dordrecht: Kluwer Academic Publishers: 61.

Coustenis, Athena. "Search for Spectroscopic Signatures of the Evaporated Atmosphere of 51 Peg b in the Near Infrared." Abstract presented at Bioastronomy 99, Hawaii, August 6, 1999.

Cowen, Ron. "Astronomers Find Two Planetary Systems." *Science News Online*, http://www.sciencenews.org (January 13, 2001).

Cowen, Ron. "Scientists Puzzle over Extra-Solar Planets." *Science News* 154, no. 6 (August 8, 1998): 88.

Croswell, Ken. "The Story of Lalande 21185." *Sky & Telescope* 89, no.6 (June 1995): 68.

Daintith, John et al. (eds.) 1994. *Biographical Encyclopedia of Scientists* 1, 2nd ed. Bristol & London: Institute of Physics Publishing.

Danner, Rudolf and Stephen Unwin (eds.). "SIM Space Interferometry Mission: Taking the Measure of the Universe." NASA JPL publication no. 400-811 (March 1999).

Davies, John K. "The Herbig Stars." *Astronomy* 13, no. 12 (December 1985): 90.

Davies, Paul. "Interplanetary Infestations." *Sky & Telescope* 98 (September 1999): 32.

Davies, Paul. "Life Force." *New Scientist* 163, 22204 (September 18, 1999): 27.

Davis, John, "Measuring the Stars," *Sky & Telescope*, October 1991, p. 361.

de Duve, Christian René. "The Beginnings of Life on Earth." *American Scientist* 83 (September–October 1995): 428.

de Duve, Christian René 1995. *Vital Dust: Life as a Cosmic Imperative*. New York: Basic Books.

Debarat, S., J. A. Eddy, H. K. Eichhorn, and A. R. Upgren (eds.) 1988. *Mapping the Sky: Past Heritage and Future Directions*. Proceedings of the 133rd Symposium of the International Astronomical Union, Paris, June 1–5, 1987. Dordrecht: Kluwer Academic Publishers.

Dick, Steven J. 1996. *The Biological Universe: The Twentieth-Century Extraterrestrial Life Debate and the Limits of Science*. New York: Cambridge University Press.

Dierickx, Philippe et al. 1999. "The Optics of the OWL 100-m Adaptive Telescope." In *Proceedings of the Backaskog Workshop on Extremely Large Telescopes*. T. Anderson, A. Ardeberg, and R. Gilmozzi (eds.): 97.

Dooling, David. "Gamma-Ray Bursts Remain as Mysterious as Ever." *Astronomy Magazine Online*, http://www.astronomy.com (October 26, 1999).

Dorminey, Bruce. "Lonely Planet." *Financial Times*, London, October 9, 1997, Technology section.

Doyle, Laurance R. "Extra-Solar Planets Around Eclipsing Binaries II: A Photometric Search of Baade's Third Window in the Galactic Plane." Abstract presented at Bioastronomy 99, Hawaii, August 6, 1999.

DuBois, Charles C. "Planets from the Very Start." *Research/Penn State* 18, no. 3. (September 1997): 29.

"Dusting Off Extra-Solar Planets." *Space Daily Express*, http://www.spacedaily.com (December 24, 1999).

"Earth Microbes on the Moon." http://www.science.nasa.gov/news (September 1, 1998).

"Eight New Very Low-Mass Companions to Solar-Type Stars Discovered at La Silla." ESO press release, no. 13/00 (May 4, 2000).

"Extra-Solar Planet in Double Star System Discovered from La Silla. Early Success with New Swiss Telescope." ESO press release, no. 18/98 (November 24, 1998).

"Extraterrestrial Civilizations: Coming of Age in the Milky Way." Space Telescope Science Institute press release PR98-43 (December 10, 1998).

Fajarda-Acosta, Sergio. "Ageing Dust Fades Away." *Nature* 401 (September 30, 1999): 439.

Favata, F. and O. Pace. "Eddington Notes." *ESA Bulletin* 105 (February 2001): 46.

"First Detection of Light from an Extra-Solar Planet." Royal Astronomical Society press release (December 15, 1999).

"First Prototype of Revolutionary SETI Telescope Unveiled at 40th Anniversary of the World's First Scientific Search for Extraterrestrial Intelligence." University of California at Berkeley press release (April 19, 2000).

"First System of Multiple Planets Around a Sun-like Star." San Francisco State University press release (April 15, 1999).

"First Visiting Astronomers at VLT Kueyen." ESO press release, no.13/00 (April 13, 2000).

Fischer, Daniel and Hilmar Duerbeck 1998. *Hubble Revisited: New Images from the Discovery Machine.* New York: Copernicus Books.

Ford, Eric B. et al. "Theoretical Implications of the PSR 1620 –26 Triple System and Its Planet." *APJ* 528, no. 336 (January 1, 2000).

Frink, Sabine et al. "Testing Hipp. K Giants as Grid Stars for SIM." Poster paper presented at "Working on the Fringe," NASA/JPL conference on interferometry, Dana Point, California, May 1999.

Frost, Pam. "Is Sol A Rare System?" *Space Daily Express*, http://www.spacedaily.com (January 13, 2000).

"Fuse Spacecraft Observes Interstellar Lifeblood of Galaxies." NASA press release (January 12, 2000).

"GAIA-Composition, Formation and Evolution of the Galaxy, Results of the Concept and Technology Study." ESTEC/ESA GAIA Science Advisory Group, draft version 1.5, (January 6, 2000).

García-Sanchez, Joan. "Stellar Encounters with the Oort Cloud Based on Hipparcos Data." *Astronomical Journal* 117 (February 1999): 1042.

Gatewood, George D. "A Study of the Astrometric Motion of Barnard's Star." *Astrophysics and Space Science* 223 (1995): 91.

Gatewood, George D. "An Astrometric Study of Lalande 21185." *Astronomical Journal* 79, no. 1 (January 1974): 52.

Gatewood, George D. "Lalande 21185." Abstract. Session 40: The Environment of Stars: From Protostars to the Main Sequence. American Astronomical Society, 188th Meeting in Madison, Wisconsin (June 11, 1996).

Gatewood, George D. "On the Astrometric Detection of Neighboring Planetary Systems." *Icarus* 27 (1976): 1.

Gatewood, George D. and Heinrich Eichorn. "An Unsuccessful Search for a Planetary Companion of Barnard's Star." *Astronomical Journal* 78, no. 8 (October 1973): 769.

Gaudi, B. Scott and colleagues at the PLANET Collaboration. "Microlensing Constraints on the Frequency of Jupiter Mass Planets." Abstract presented at the American Astronomical Society Annual Meeting, Atlanta, Georgia, January 2000.

Gliese, Wilhelm 1969. *Catalogue of Nearby Stars.* Karlsruhe: G. Braun.

Gliese, Wilhelm 1976. "Preliminary List of Star Catalogues 1963–1976." Heidelberg: Astronomisches Rechen-Institut.

Gliese, Wilhelm 1963. *The Right Ascension System of the Fourth Fundamental Catalogue (FK4).* Karlsruhe: G. Braun.

Gliese, Wilhelm, C. Andrew Murray, and R. H. Tucker (eds.) 1974. *New Problems in Astrometry.* Boston/Dordrecht: Reidel.

Goldin, Dan. "Speech Welcoming New Director of NASA's Astrobiology Institute." Ames Astrobiology Institute, May 18, 1999.

Goodwin, S. P. et al. "The Relative Size of the Milky Way." *The Observatory* 118, no. 1145 (August 1998): 201.

Guillermier, Pierre and Serge Koutchmy 1999. *Total Eclipses: Science, Observations, Myths and Legends.* London: Springer-Verlag Praxis Series.

Habing, H. J. "Disappearance of Stellar Debris Disks Around Main-Sequence Stars after 400 Million Years." *Nature* 401 (September 30, 1999): 456.

Han, Inwoo, David C. Black, and George D. Gatewood. "Preliminary Astrometric Masses for Proposed Extra-Solar Planetary Companions." Paper presented at the 2000 Meeting of the Division for Planetary Sciences of the American Astronomical Society, Pasadena, California, October 23–27, 2000.

Hart, Michael H. 1982. "An Explanation for the Absence of Extraterrestrials on Earth." In *Extraterrestrials—Where Are They?* Michael Hart and Ben Zuckerman (eds.). Elmsford, U.K.: Pergamon Press: 2.

Hayden, Thomas. "Curtain Call." *Astronomy* 28 (January 2000): 45.

"HD 209458: When Hipparcos Saw the Shadow of an Alien Planet." *ESA Science News* (December 14, 1999).

Heacox, William. "On the Nature of Sub-Stellar Mass Companions to Solar-Like Stars." Paper presented at Bioastronomy 99, Hawaii, August 6, 1999.

Henahan, Sean. "From Primordial Soup to the Prebiotic Beach: An Interview with Exobiology Pioneer Dr. Stanley L. Miller University of California San Diego." *Access Excellence* website, http://www.accessex-cellence.com/WN/NM/miller.htm (March 2000).

"Hints of Planet-Sized Objects Bewilder Hubble Scientists." JPL press release (June 27, 2001).

Hinz, Philip M. et al. "Imaging Circumstellar Environments with a Nulling Interferometer." *Nature* 395 (September 17, 1998): 251.

"Hitchhiking Molecules Could Have Survived Comet Collision with Earth." University of California at Berkeley press release (April 4, 2001).

Holland, Wayne S. et al. "Submillimetre Images of Dusty Debris Around Nearby Stars." *Nature* 392 (1998): 788.

Howell, Steve. "Wyoming-Arizona Search for Planets (WASP)." WASP home page (1999).

"Hubble Resumes Gazing at the Heavens by Taking a Look at the 'Eskimo' Nebula." Space Telescope Science Institute press release (January 24, 2000).

"Hubble Space Telescope Captures First Direct Image of a Star." STScI press release (January 15, 1996).

"Hubble Spys Giant Elliptical Ring System in Beta Pictoris." *Space Daily Express*, http://www.spacedaily.com (January 17, 2000).

"Hubble Surveys Dying Suns in Nearby Galaxy." Space Telescope Science Institute press release (March 9, 2000).

Hughes, John P. et al. "Nucleosynthesis and Mixing in Cassiopeia A." *The Astrophysical Journal* 528 (January 10, 2000): L109.

"ISO Measures Possible Planetary System in Formation." *ESA Science News* (April 20, 2000).

Jayawardhana, Ray et al. "A Dust Disk Surrounding the Young A Star HR 4796A." *The Astrophysical Journal Letters* 503 (August 10, 1998): L79–L82.

Jeans, James. "Is There Life on the Other Worlds?" *Science* 95 (June 12, 1942): 589.

Kaler, James B. 2001. *The Little Book of Stars*. New York: Copernicus Books.

Kasting, James F. "Earth's Early Atmosphere." *Science* 259 (February 12, 1993): 920.

Kasting, James F. et al. "Habitable Zones Around Main Sequence Stars." *Icarus* 101 (1993): 108.

Kasting, James F. and Darren M. Williams. "Habitable Planets with High Obliquities." *Icarus* 129 (1997): 254.

Kelleher, Florence M. "Edward Emerson Barnard." http://astro.uchicago.edu/yerkes/virtualmuseum/Barnardfull.html (April 2, 1997).

Kessler, M. F. et al. "ISO's Astronomical Harvest Continues." *ESA Bulletin* (September 1999): 6.

Koerner, W. et al. "Stars and Disks: A Single Circumbinary Disk in the HD 98800 Quadruple System." Abstract from 2000 American Astronomical Society Annual Meeting, Atlanta, Georgia. January 11–15, 2000.

Konacki, Maciej, doctoral candidate in astronomy at Nicolaus Copernicus University in Torun, Poland. Interviewed on September 1, 1999, at Bonn, Germany.

Labeyrie, Antoine 1999. "Direct Searches: Imaging, Dark Speckle and Coronography." In *Planets Outside the Solar System: Theory and Observations*. J. M. Mariotti and D. Alloin (eds.). Dordrecht: Kluwer Academic Publishers: 261.

Labeyrie, Antoine. "Kilometric Arrays of 27 Telescopes: Studies and Prototyping for Elements of 0.2 m, 1.5 m, and 12–25 m Size." *SPIE* 3350, OHP preprint 114.

Labeyrie, Antoine. "Snapshots of Alien Worlds—The Future of Interferometry." *Science* 285 (September 17, 1999): 1864.

Lacy, C. "Absolute Dimensions and Masses of the Remarkable Spotted dM4e Eclipsing Binary Flare Star CM Draconis." *The Astrophysical Journal* 218 (December 1977): 444.

Laskar, Jacques and P. Robutel. "The Chaotic Obliquity of the Planets." *Nature* 361 (February 18, 1993): 608.

Laskar, Jacques, P. Robutel, and F. Joutel. "Stabilization of the Earth's Obliquity by the Moon." *Nature* 361 (February 18, 1993): 615.

Laskar, Jacques. "The Moon and the Origin of Life on Earth." Private translation of paper originally published in *Pour la Science* (April 1993).

Leger, A. et al. 1994. "How to Search for Extra-Solar Planets with the VLT/VISA." In *Science with the VLT—Proceedings of the ESO Workshop at Garching Walsh.* Jeremy R. Walsh and Ivan J. Danziger (eds.). Germany: Springer-Verlag: 21.

Leonard, Peter J. T. and Jerry T. Bonnell. "Gamma-Ray Bursts of Doom." *Sky & Telescope* (February 1998): 28.

Leonard, Peter, astronomer at NASA Goddard Space Flight Center in Greenbelt, Maryland. Private communication with author on November 1, 1999.

Leverington, David 1995. *A History of Astronomy: From 1890 to the Present.* London: Springer-Verlag.

Lifton, Sarah 1999. *Perspectives.* SETI Institute Brochure/Publication.

Lineweaver, Charles H. "An Estimate of the Age Distribution of Terrestrial Planets in the Universe: Quantifying Metallicity as a Selection Effect." *Icarus* 151 (2001): 307.

Lineweaver, Charles H. "Why Did We Evolve So Late?" Poster paper presented at Bioastronomy 99, Hawaii, August 6, 1999.

Lippincott, Sarah Lee. "Astrometric Analysis of Lalande 21185." *Astronomical Journal* 65, no. 7 (September 1960): 445.

Liseau, R. "Molecular Line Observations of Southern Main Sequence Stars with Dust Disks: Alpha Ps A, Beta Pic, Epsilon Eridania, HR 4796A." *Astronomy & Astrophysics* 348 (1998): 133.

Lissauer, Jack J. "Three Planets for Upsilon Andromedae." *Nature* 398 (April 22, 1999): 659.

Livio, Mario. "How Rare Are Extraterrestrial Civilizations, and When Did They Emerge?" *The Astrophysical Journal* 511 (January 20, 1999): 429.

Livio, Mario and Lionel Siess. "The Accretion of Brown Dwarfs and Planets by Giant Stars—II. Solar-Mass Stars on the Red Giant Branch." *Monthly Notices of the Royal Astronomical Society* 308 (1999): 1133.

"Lunar Prospector Magnetic Data Supports Popular Theory of Unique Moon Formation." *Space Daily Express,* http://www.spacedaily.com (August 9, 1999).

Lunine, Jonathan I. "The Occurrence of Jovian Planets and the Habitability of Planetary Systems." *Proceedings of the National Academy of Sciences* 98, no. 3 (January 30, 2001): 809.

"M22." Students for the Exploration and Development of Space (SEDS) chapter at the University of Arizona's Lunar and Planetary Laboratory. http://www.seds.lpl.arizona.edu/messier (December 9, 1999).

Maccone, Claudio 1999. "Space Missions Enabling SETI Searches Farther and Farther Out." http://www.nidisci.org.essaycomp/cmaccone.

MacDermott, Alexandra. "Chirality." *New Scientist, RBI Limited 2000,* http://www.newscientist.com/nspplus/insight/future/macdermott.html

MacDermott, Alexandra 1997. "Distinguishing the Chiral Signature of Life in the Solar System and Beyond." *SPIE* 3111: 272.

Malhotra, Renu. "Chaotic Planet Formation." *Nature* 402 (December 9, 1999): 599.

Marcy, Geoffrey and Paul Butler 1999. "Distribution of Masses: Planets and Brown Dwarfs." http://www.physics.sfsu.edu/~gmarcy/planetsearc/planetsearch.html

Marcy, Geoffrey and Paul Butler. "Hunting Planets Beyond." *Astronomy* 28, no. 3 (March 2000): 43.

Marley, Mark S. et al. "Reflected Spectra and Albedos of Extra-Solar Giant Planets. I. Clear and Cloudy Atmospheres." *The Astrophysical Journal* 513: 879.

Marley, Mark S. 1998. "Ground- and Space-Based Spectroscopy of the Composition of Brown Dwarfs and Extra-Solar Planets." In *Failed Stars and Super Planets: A Report Based on the January 1998 Workshop on Substellar-Mass Objects.* Space Studies Board, National Research Council. Washington, D.C.: National Academy Press: 22.

"Marshall Scoping Exo Worlds." *Space Daily Express*, http://www.spacedaily.com (October 13, 1999).

Mason, Brian D. and William I. Hartkopf. "The New WDS Designation Scheme." *The Washington Double Star Catalog.* U.S. Naval Observatory, Washington, D.C. Posted at http://www.iau.org in August 2000.

"Martian Meteorites Reveal Clues to Processes in Planet's Atmosphere." University of California at San Diego press release (March 1, 2000).

Massey, Philip and Michael R. Meyer 2001. "Stellar Masses." *Encyclopedia of Astronomy and Astrophysics* 4 Bristol & Philadelphia: IOP and Macmillan Publishers Ltd.: 3103.

Mather, John C. 1997. "Planets Move Beyond the Solar System and the Next Generation of Space Missions." In *ASP Conference*, Series 119. Astronomical Society of the Pacific. D. R. Soderblom (ed.): 245.

Mathewson, D. "The Clouds of Magellan," *Scientific American* 252 (April 1985): 106.

McBreen, Brian and L. Hanlon. "Gamma-Ray Burst and the Origin of Chondrules and Planets." *Astronomy & Astrophysics* 351 (1999): 759.

Ménard, François et al."High Chirality Polarisation in the Star Formation Region NGC 6334: Implication for Biomolecular Homochirality." Abstract presented at Bioastronomy 99, Hawaii August 6, 1999.

Miller, Stanley L. and Harold C. Urey. "Organic Compound Synthesis on the Primitive Earth," *Science* 130, no. 3370 (July 31, 1959): 245.

Moomaw, Bruce. "Orbital Resonance and Smashing Planets Not All Habitable Zones Are Created Equal." *Space Daily Express*, http://www.spacedaily.com (August 9, 1999).

"Most Distant Spacecraft May Reach Shock Zone Soon." NASA JPL press release (December 18, 2000).

Naeye, Robert. "Transit Planets." *Astronomy* (March 2000): 46.

"NASA Establishes Contact with Famed Pioneer 10 Spacecraft." NASA Ames Research Center, Moffett Field, California, press release (April 30, 2001).

"NASA Gives Official Nod to First Mercury Orbiter Mission." NASA press release (June 7, 2001).

"NASA Scientists Find Clues That Life Began in Deep Space." NASA Ames Research Center press release (January 26, 2001).

"NASA Selects Two Investigations for Pluto-Kuiper Belt Mission Feasibility Studies." JPL/Caltech/NASA media release (June 6, 2001).

"NASA's STARDUST Spacecraft, Bound for Comet Wild-2, Celebrated Its First Year in Space This Month." *NASA Space Science News* (February 27, 2000).

"NEAR Shoemaker Observations Link Eros to Primordial Solar System." Johns Hopkins University Applied Physics Laboratory press release (May 30, 2000).

"New Theory Links Moon's Current Orbit to Its Formation via a Giant Impact." Southwest Research Institute, San Antonio, Texas, press release (February 16, 2000).

"Newly Discovered Bright Supernova Is Testament to Value of UC Berkeley's Robotic Telescope and the Only Fully Automated Supernova Search." University of California at Berkeley press release (November 11, 1999).

Norris Russell, Henry. "Physical Characteristics of Stellar Companions of Small Mass." *Publications of the Astronomical Society of the Pacific* 55 (April 1943): 79.

Norris, Ray P. "How Old Is ET?" Paper presented at Bioastronomy 99, Hawaii, August 6, 1999.

North, Gerald 1997. *Astronomy Explained*. London: Springer-Verlag.

North, John 1994. *The Fontana History of Astronomy & Cosmology*. London: Fontana Press.

Noyes, Robert et al. "A Planet Orbiting the Star Rho Coronae Borealis." *The Astrophysical Journal* 48 (July 10, 1997): L111.

"On Designating Components of Binary / Multiple Star Systems: A Multi-Commission Meeting on Designations of Stellar Components: IAU Resolution of C Type for Commissions 26 and 42." Sponsored by Commissions 5 and 26, the XXIVth General Assembly, Manchester, U.K. (August 2000).

"One Theory Solves Two Ancient Climate Paradoxes." Penn State press release (December 14, 1999).

Oppenheimer, Ben R. 2001. "Friedrich Bessel and the Companion of Sirius." In *Cosmic Horizons: Astronomy at the Cutting Edge*. Steven Soter and Neill de Grasse Tyson (eds.). New York: The New Press: 67.

Oppenheimer, Ben R., University of California at Berkeley. Private communication with author on June 12, 2001.

Oort, Jan H. "The Formation of Galaxies and the Origin of the High-Velocity Hydrogen." *Astronomy & Astrophysics* 7 (September 1970): 381.

Orgel, Leslie. "The Origin of Life on Earth." *Scientific American* 271, no. 4 (October 1994): 76.

Owen, Tobias et al. "A Low-Temperature Origin for the Planetesimals That Formed Jupiter." *Nature* 402 (November 18, 1999): 269.

Pacchioli, David. "Worlds Beyond the Sun." *Research/Penn State* 18, no. 1 (January 1997): 14.

Padgett, Deborah L. et al. "Hubble Space Telescope/NICMOS Imaging of Disks and Envelopes around Very Young Stars." *Astronomical Journal* 117 (March 1999): 1490.

Perryman, Michael. "A Stereoscopic View of Our Galaxy." *Astronomy & Astrophysics* 40, no. 6 (December 1999): 23.

Perryman, Michael. "Hipparcos: The Stars in Three Dimensions." *Sky & Telescope* 97 (June 1999): 40.

Peterson, Charles J. 1997. "Magellanic Clouds." In *History of Astronomy: An Encyclopedia*. John Lankford (ed.). New York: Garland Publishing, Inc.: 317.

"Physics News Update." *Bulletin of Physics News*, no. 520 (January 12, 2001). The American Institute of Physics.

"Planetary Society Turns Eyes to the Skies for ET." Planetary Society press release (January 19, 1999).

"Planets, Planets Everywhere." *ESA Bulletin* 100 (December 1999): 106.

Pravdo, S. and S. Shaklan. "Astrometric Detection of Extra-Solar Planets: Results of a Feasibility Study with the Palomar 5-M." *Astrophysical Journal* 465, no. 1 (July 1, 1996): 264.

"Project Phoenix Observations: Minus 3." SETI Institute website (October 29, 1999).

Queloz, Didier 1999. "Indirect Searches: Doppler Spectroscopy & Pulsar Timing." In *Planets Outside the Solar System: Theory and Observations*. J. M. Mariotti and D. Alloin (eds.). Dordrecht: Kluwer Academic Publishers: 229.

Queloz, Didier 1999. "The New Planetary Systems." In *Planets Outside the Solar System: Theory and Observations*. J. M. Mariotti and D. Alloin (eds.). Dordrecht: Kluwer Academic Publishers: 107.

Quirrenbach, A. 1994. "Astrometric Detection and Investigation of Planetary Systems with the VLT Interferometer." In *Science with the VLT—Proceedings of the ESO Workshop at Garching, Germany*. Jeremy R. Walsh and Ivan J. Danziger (eds.). Berlin: Springer-Verlag: 33.

"Rare Meteorite Considered Most Valuable Find in Decades." Reuters News Service (June 1, 2000).

Rees, Martin. "Exploring Our Universe and Others." *Scientific American* 281, no. 6 (December 1999): 44.

"Researchers Discover Extraterrestrial Gases in Buckyballs." NASA Ames press release (March 20, 2000).

Reuyl, Dirk and Erik Holmberg. "On the Existence of a Third Component in the System 70 Ophiuchi." *The Astrophysical Journal* 97 (January–May 1943): 41.

"Robotic Telescope Captures Visible Light from a Powerful Gamma-Ray Burst." *NASA Science News* (January 27, 1999).

Rodriquez, L. F. "Compact Protoplanetary Disks Around the Stars of a Young Binary System." *Nature* 395 (September 24, 1998): 355.

Rottler, Lee. "Extra-Solar Planets Around Eclipsing Binaries II: A Photometric Search in the Globular Cluster NGC 6752." Abstract presented at Bioastronomy 99, Hawaii, August 6, 1999.

Sackett, Penny D. "A Search for Other Planetary Systems: Stars and Dark Objects as Microlenses in the Milky Way." Kapteyn Astronomical Institute, Rijksuniversiteit Groningen, *PLANET Online*, http://www.astro.rug.nl/~psackett/NVWS/MicroPLANET.html

Sackett, Penny D. 1999. "Searching for Unseen Planets via Occultation and Microlensing." In *Planets Outside the Solar System: Theory and Observations*. J. M. Mariotti and D. Alloin (eds.). Dordrecht: Kluwer Academic Publishers: 189.

Sagan, Carl et al. "A Search for Life on Earth from the Galileo Spacecraft." *Nature* 365 (October 21, 1993): 715.

Sahai, Raghvendra et al. "Detection and Characterization of Nearby Giant Planets and Brown Dwarf Companions with an NGST Coronagraph." Abstract presented at "From Giant Planets to Cool Stars Conference", Flagstaff, Arizona, June 1999.

Sahu, Kailash C. et al."Gravitational Microlensing by Low-Mass Objects in the Globular Cluster M22." *Nature* 411 (June 28, 2001): 1022.

"SALT Telescope." South African Astronomical Observatory press release, Cape Town, (November 25, 1999).

Sanders, Robert. "Astronomers See Shadow of Planet Cross Distant Star, Proving That Extra-Solar Planets Are Real." University of California at Berkeley press release (November 12, 1999).

Sawyer, Kathy. "Astronomers Find Two New Planets: Young British Student Contributes to Discoveries Outside Solar System." *Washington Post*, September 24, 1998.

Schilling, Govert. "Giant Eyes on the Sky." *Astronomy* (December 1999): 49.

Schilling, Govert. "Jan Oort Remembered." *Sky & Telescope* (April 1993): 44.

Schilling, Govert. "Peter van de Kamp and His 'Lovely Barnard's Star.'" *Astronomy* (December 1985): 26.

"Science with the GAIA Mission." http://sci.esa.int/home/gaia/index.cfm.

"Scientist and Colleagues Show Fullerenes Can Be Cosmic Carriers." University of Hawaii press release (March 20, 2000).

"Search for Extra-Solar Planets Hits Home." McDonald Observatory of the University of Texas at Austin press release (August 7, 2000).

Seife, Charles. "Let's Learn Lincos." *New Scientist* (September 18, 1999): 36.

"SETI@Home Achieves 2 Million Mark by First Birthday." Planetary Society press release (May 16, 2000).

Setterfield, Barry. "The Vacuum, Light Speed, and the Redshift." Revisions posted November 15, 1999. Editions December 14, 1999, January 10, 2000, and February 18, 2000, http://www.ldolphin.org/setterfield/vacuum.html.

"Short and Long Gamma-Ray Bursts Are of Different Origin." NASA press release 00-132 (November 7, 2000.)

Shostak, Seth. "A New Telescope for SETI." *SETI News* 1st Quarter (1999): 5.

Shostak, Seth. "Astronomer Seth Shostak Returns to Arecibo to Listen for E.T." *abcNEWS.com* (February 23–29, 2000): Parts 1–IV.

Siegfried, Tom. "Exodus from Earth." *Astronomy* (January 2000): 51.

"Sizzling Comets Circle a Dying Star." *NASA Science News* (July 11, 2001).

"SOFIA." NASA Ames press release (November 2, 1999).

"Some New Planets Might Be Stars, Researchers Say." University of Pittsburgh news release (October 28, 2000).

"South African Crater." Agence France Presse, Johannesburg (October 5, 1999).

"Stardust Detects Organic Molecules." *Space Daily Express*, http://www.spacedaily.com (April 27, 2000).

"Stars & Planetary Systems." *Gemini Newsletter* (June 1997).

"Stellar Apocalypse Yields First Evidence of Water-Bearing Worlds Beyond Our Solar System." NASA press release (July 11, 2001).

"Stellar Wind Gap Points to Cold Distant Worlds." *Space Daily Express*, http://www.spacedaily.com (January 13, 2000).

"Steward Observatory Astronomers Head Two SIRTF Science Teams." University of Arizona news release (November 21, 2000).

Stiles, Lori. "Steward Mirror Lab to Cast 2nd 8.4-meter LBT Mirror." University of Arizona News Service (undated).

Stix, Gary. "A New Eye Opens on the Cosmos." *Scientific American* 280, no. 4 (April 1999): 104.

Strand, Kaj A. "61 Cygni as a Triple System." *Publications of the Astronomical Society of the Pacific* 55 (1943): 29.

Strand, Kaj A. "The Double Star 61 Cygni." *Sky & Telescope* (January 1942): 6.

"Strange 'Spin Cycle' Inside the Sun May Explain Sunspots, Solar Flares, and Other Mysteries." Stanford University press release (undated).

Strauss, Michael A. et al. 1999. "The Discovery of a Field Methane Dwarf from Sloan Digital Sky Survey Commissioning Data." *The Astrophysical Journal Letters* 522, no. 61.

"Students Prepare Hyperspectral Imaging Package for NASA Launch—Imager Could Eventually Reveal Life Beyond the Solar System." Washington University at St. Louis press release (undated).

"Students Use VLA to Make Startling Brown Dwarf Discovery." NRAO press release (March 14, 2001).

"Successful First Light for VLT High-Resolution Spectrograph." ESO press release 15/99 (October 5, 1999).

"Supernova Explosions May Create Star Forming Clouds." *Astronomy Magazine Online*, http://www.astronomy.com (February 21, 2000).

Tarter, Jill C. and Christopher F. Chyba. "Is There Life Elsewhere in the Universe?" *Scientific American* (December 1999): 80.

Tattersall, Ian. "Once We Were Not Alone." *Scientific American* (January 2000): 39.

Telesco, C. M. "Deep 10 and 18 Micrometer Imaging of the HR 4796A Circumstellar Disk: Transient Dust Particles & Tentative Evidence for a Brightness Asymmetry." *The Astrophysical Journal* 329: 329.

"The Crystalline Revolution: ISO's Finding Opens a New Research Field, 'Astro-Mineralogy.'" *ESA Science News* (February 4, 2000).

"The Shifting Stars That Made Einstein Famous." *ESA Science News* (July 29, 1999).

"The Stuff Between the Stars." *NASA Science News* (July 31, 2000).

"The Telescope, the Observations, and the Servicing Mission." ESA press release (November 24, 1999).

Thommes, Edward W. et al. "The Formation of Uranus and Neptune in the Jupiter-Saturn Region of the Solar System." *Nature* 402 (December 9, 1999): 635.

Transit of Extra-Solar Planets (TEP) home page, http://www.mit.edu/~tep (1999).

Trilling, D. E. and R. H. Brown. "Imaging a Circumstellar Disk Around a Star with a Radial-Velocity Planetary Companion." Paper presented at the AAS DPS meeting, August 11, 1998.

Trimble, Virginia 1995. "Galactic Chemical Evolution: Implications for the Existence of Habitable Planets." In *Extraterrestrials: Where Are They?* 2nd ed. Ben Zuckerman and Michael H. Hart (eds.). Cambridge, U.K.: Cambridge University Press: 184.

Turon, Catherine. "From Hipparchus to Hipparcos: Measuring the Universe, One Star at a Time." Adapted from *Sky & Telescope Online* (1999).

"UA Scientists Are First to Discover Debris Disk Around Star Orbited by a Planet." University of Arizona News Service (1998).

"UCLA Astronomers Identify Evidence of Asteroid Belt Around Nearby Star: Findings Indicate Potential for Planet or Asteroid Formation." UCLA press release (June 4, 2001).

Unwin, Stephen. "SIM Proceeds toward Phase B." In *Fringes*. Stephen Unwin (ed.). Space Interferometry Mission newsletter, no. 16 (May 23, 2001).

"U.S. Astronomers Meet at Arecibo to Discuss Spending, Siting and Science for Mammoth-Size, Next-Generation Radio Telescope." Cornell University press release (March 3, 2000).

"U.S. Naval Observatory Satellite to Measure Positions of 40 Million Stars." U.S. Naval Observatory press release (October 15, 1999).

"U.S., European ALMA Partners Award Prototype Antenna Contracts." National Radio Astronomy Observatory press release (March 14, 2000).

Valtonen, M. J. and J. Q. Zheng. "Transfer of Potentially Life-Carrying Meteoroids from One Planetary System to Another." Abstract from the 2000 American Astronomical Society Annual Meeting, Atlanta, Georgia, January 11–15, 2000.

van de Kamp, Peter. "Alternate Dynamical Analysis of Barnard's Star." *Astronomical Journal* 74, no. 6 (August 1969): 757.

van de Kamp, Peter. "Astrometric Study of Barnard's Star from Plates Taken with the 24-inch Sproul Refractor." *Astronomical Journal* 68, no. 7 (September 1963): 515.

van de Kamp, Peter. "Astrometric Study of Barnard's Star from Plates Taken with the Sproul 61-cm Refractor." *Astronomical Journal* 80, no. 8 (August 1975): 658.

van de Kamp, Peter. "Barnard's Star as an Astrometric Binary." *Sky & Telescope* (July 1963): 8.

van de Kamp, Peter. "Parallax, Proper Motion, Acceleration, and Orbital Motion of Barnard's Star." *Astronomical Journal* 74, no. 2 (March 1969): 238.

Vidal-Madjar, A. et al. "Beta Pictoris, a Young Planetary System? A Review." *Planetary Space Science* 46, no. 6/7 (June 1998): 629.

Wade, Nicholas. "The Origins of Life Get Murkier." *International Herald Tribune* (June 15, 2000). First published in the Science section of the *New York Times*, June 13, 2000.

Walker, Gordon A. H. "A Search for Jupiter-Mass Companions to Nearby Stars." *Icarus* 116 (1995): 359.

Walker, Gordon A. H. "A Solar System Next Door." *Nature* 382 (July 4, 1996): 23.

Walker, Gordon A. H. et al. "Gamma Cephei: Rotation or Planetary Companion." *The Astrophysical Journal* 396 (September 10, 1992): L91.

"Wandering Mystery Planets." *NASA Science News* (June 29, 2001).

Warren Jr., Wayne H. "Proposal for a System of Nomenclature for Double and Multiple Objects Outside the Solar System." http://www.iau.org.

Welther, Barbara L. "The Impact of the Henry Draper Catalog on 20th-Century Astronomy." Abstract from the 195th American Astronomical Society Meeting, Atlanta, Georgia, January 2000.

"When Is a Planet Not a Planet?" University of Pittsburgh press release (October 25, 2000).

Wilford, John Noble. "Found: 2 Planetary Systems. Result: Astronomers Stunned." *New York Times* (January 10, 2001).

Wilford, John Noble. "Student's Tip Helps to Find New Planets Beyond Sun." *New York Times* (September 24, 1998).

Williams, Darren M. et al. "Habitable Moons Around Extra-Solar Giant Planets." *Nature* vol. 385 (January 16, 1997): 234.

"With the Discovery of Extra-Solar Planets Smaller Than Saturn Astronomers Are Increasingly Convinced That Other Stars Harbor Planetary Systems Like Our Own." *NASA Science News* (March 29, 2000).

"With the VLT Interferometer Towards Sharper Vision." ESO press release (May 24, 2000).

Wolstencroft, Ramon D. and John A. Raven. "Viability and Detectability of Photosynthesis on Earth-Like Planets Orbiting Main Sequence Stars." Paper presented at Bioastronomy 99, Hawaii, August 6, 1999.

Wolszczan, Alexander. "New Observations of Pulsar Planets." Abstract from Pulsar 2000 Colloquium, Bonn, Germany, August 30–September 3, 1999.

Wong, Janet. "Jupiter: Destroyer of Worlds," *Space Daily Express*, http://www.spacedaily.com (December 8, 1999).

"X-Ray Star Stuff." *NASA Science News* (July 18, 2000).

"Yukon Meteor Blast." *NASA Science News* (January 25, 2000).

"Yukon Meteorite Recovered." *NASA Science News* (March 16, 2000).

Zabarenko, Deborah. "Astronomers Find Two New Planetary Systems." Reuters News Service (January 10, 2001).

Zapatero Osorio, Maria Rosa et al. "Discovery of Young, Isolated Planetary Mass Objects in the Sigma Orionis Star Cluster." *Science* 290 (October 6, 2000): 103.

Index

Page references to illustrations are in italics.

A

AAA. *See* Absolute Astronomical Accelerometer

AAS. *See* American Astronomical Society

Abell, George, 16, 148

Abell 39 (nebula), *16*

Absolute Astronomical Accelerometer (AAA), 66

absorption lines, 37, 40

acids, 164–65

Akeson, Rachel, 130

alanine, *164*

Allegheny Observatory, 115–*17*, 191

Allen, Paul, 182

Allen Telescope Array, 182–*83*

ALMA. *See* Atacama Large Millimeter Array

Almagest (Ptolemy), 109

Alpha Centauri, 152

Alpha Orionis. *See* Betelgeuse

Alpha Ursae Minoris, 98

aluminum, 29

Amalthea (moon of Jupiter), 111

American Astronomical Society (AAS), 14, 76, 86, 116, 191

Ames Research Center, 114, 159

amino acids, *164*–65, 171

ANDICAM. *See* Novel Double-Imaging CAMera, A

Angel, Roger, 137–39, 175

Anglo-Australian Observatory, 166

angular momentum, 4, 22–23

antennas, parabolic, 30

Antofagasta, Chile, 143–44, 152

Apps, Kevin, 56–57

Arcturus (star), 109

Arecibo Observatory, 181

pulsar studies at, *11*–12, 14

argon, 72

asteroids, 27, 87, 155, 167, 170

asteroseismology, 92, 127

astrobiology, 174–76

astrometric tiles, 135

astrometry, 109–19, 128, 135, 156, 191

astronomers

anticipation of data, 30, 32, 65, 70, 93

patience required of, 2, 17, 36, 62–63, 66, 77, 116

Astronomical Journal, The, 55

Astronomical Multiple Beam Recombiner (AMBER), 150

astronomical unit (AU)

defined, 27

astronomy

astrometry, 109–19, 128, 135, 156, 191

interferometry, 105, 107, *122*–23, 124–*26*, 127–*31*, 132–40, 135, 149–*51*, 158, 175

occultation, 87, 109, 118

pulsar, 7–17, 22, 36, 52, 95, 191

radio, 5, 9–*11*, 9–16, 12–13, *14*–16, 32, 61, 92–93, 180–85

and pulsars, 9–17

signal interference, 184

spectroscopic, 36–50, 52, 57–58, 62, 66–70, 75, 93, 109–10, 116, 119, 130, 134, 175, 186, passim

See also Earth; planets; Solar System; stars; Sun; telescopes

Astronomy, 114

Atacama Large Millimeter Array (ALMA), *30*

atmosphere, of Earth, 11

absorption of gamma rays, 28

effect on star coloration, 42–43

effects on telescopic work, 21–22, 127, 130, 155

hydrogen lines in, 44–45

modeling of, 128, 130

Atomic Energy Commission, U.S., 28

atomic weight, 2

Atwood, Bruce, *83*

Austin, Texas, 58–59

B

Baade's Third Window, *90*

Babcock, Horace, 128–29

Backer, Donald, 15

bacteria, 172–73

Bailey, Jeremy, 167

Baltimore, Maryland, 91, 159

Barnard, Edward Emerson, 111–*12*

Barnard's Star, 111–*12*, 113–14, 134

Bean, Alan, 173

Bell Burnell, Jocelyn, 9–11, 16–17

Berkeley, California, 54

Bernstein, Max, 165–66

Beta Pictoris, *26*–27, 31

Betelgeuse (star), 105, 124–25, 127, 130

nulling interferometry, 130

Bethe, Hans, 41